SIGNIFICANT CHANGES TO THE
INTERNATIONAL RESIDENTIAL CODE

2006 EDITION

THOMSON

DELMAR LEARNING ™

Australia Canada Mexico Singapore Spain United Kingdom United States

THOMSON

DELMAR LEARNING

Significant Changes to the International Residential Code 2006 Edition
Hamid A. Naderi and Douglas W. Thornburg

Vice President, Technology Professional Business Unit:
Gregory L. Clayton

Product Development Manager:
Ed Francis

Director of Marketing:
Beth A. Lutz

Development:
Ohlinger Publishing Services

Marketing Coordinator:
Marissa Maiella

Production Director:
Patty Stephan

Production Manager:
Andrew Crouth

Content Project Manager:
Kara A. DiCaterino

Art and Design Coordinator:
Francis Hogan

Editorial Assistant:
Sarah Boone

Library of Congress Cataloging-in-Publication Data:
Card Number:
Naderi, Hamid.
    Significant changes to the International residential code 2006 edition / Hamid Naderi, Doug Thornburg.
       p. cm.
    ISBN: 1-4180-2878-9
1.   Building—Standards. 2.   Construction industry—Standards.   I. Thornburg, Doug. II. Title.
TH420.N32 2006
690'.8370218—dc22

                                    2006044486

ISBN: 1-4180-2878-9

NOTICE TO THE READER

Contents

PART 1
Administration (IRC Chapter 1) 1

■ **R101.2**
Scope 2

■ **R108.2**
Schedule of Permit Fees 4

■ **R110.1**
Use and Occupancy 5

PART 2
Definitions (IRC Chapter 2) 7

■ **R202 Definitions**
Accessory Structure 8

■ **R202 Definitions**
Approved 10

■ **R202 Definitions**
Exterior Wall 11

■ **R202 Definitions**
Fire-Separation Distance 13

PART 3
Building Planning and Construction (IRC Chapters 3–10) 16

■ **R301.2.1.1**
Design Criteria 20

■ **R301.2.1.2 and Table R301.2.1.2**
Protection of Openings 22

■ **R301.2.2, Table R301.2.2.1.1, and Figure R301.2(2)**
Seismic Provisions 25

■ **R301.2.2.2.2**
Irregular Buildings 27

■ **Table R301.5**
Minimum Uniformly Distributed Live Loads 29

■ **R302**
Exterior Wall and Opening Protection 32

■ **R303.6.1**
Light Activation at Stairways 35

■ **R305.1**
Minimum Height of Sloped Ceilings 37

■ **R308.3**
Glazing Materials Permitted in Hazardous Locations 39

■ **R308.4**
Glazing Adjacent to Stairways and Landings 41

■ **R309.2**
Separation of Detached Garage from Dwelling 43

■ **R310.1**
Emergency Escape and Rescue Openings 45

■ **R310.1.4, R310.4**
Operation of Emergency Escape
and Rescue Openings — 47

■ **R310.5**
Emergency Openings under Decks
and Porches — 49

■ **R311.4.3**
Landings at Exterior Doors — 51

■ **R311.5.4**
Landings at Garage Stairways — 53

■ **R311.6.1**
Maximum Slope of Ramps — 55

■ **R312.1**
Guards at Elevated Ramps — 57

■ **R313.1**
Smoke Alarms and Household Fire
Alarm Systems — 58

■ **R313.1.1**
Smoke Alarms in Existing Dwellings — 60

■ **R317.1**
Fire Separation of Two-Family Dwellings — 61

■ **R319.1, R202**
Protection of Wood Members
against Decay — 63

■ **R319.1.5**
Protection of Glued-Laminated
Members against Decay — 65

■ **R320, Table R301.2(1)**
Protection against Subterranean Termites — 66

■ **R324.1.3.1**
Determination of Design Flood Elevations — 69

■ **R401.3**
Drainage — 71

■ **R401.4.2**
Compressive or Shifting Soil — 73

■ **Table R402.2**
Minimum Specified Compressive
Strength of Concrete — 75

■ **R403.1.4.2**
Seismic Conditions — 77

■ **R403.1.6.1**
Foundation Anchorage in Seismic Design
Categories C, D_0, D_1, and D_2 — 78

■ **R404.1**
Concrete and Masonry Foundation Walls — 80

■ **R404.5**
Retaining Walls — 84

■ **R406.1**
Concrete and Masonry Foundation
Dampproofing — 85

■ **R406.2**
Concrete and Masonry Foundation
Waterproofing — 87

■ **R408.2, R408.3**
Openings for Under-Floor Ventilation
(R408.2) and Unvented Crawl Space (R408.3) — 89

■ **R502.2.1 and R602.10.8**
Framing at Braced Wall Lines (R502.2.1)
and Connections (R602.10.8) — 92

■ **Table R502.5(1)**
Girder Spans and Header Spans
for Exterior Bearing Walls — 94

■ **R502.13**
Fireblocking Required — 97

■ **Table R503.2.1.1(1)**
Allowable Spans and Loads for Wood
Structural Panels for Roof and Subfloor
Sheathing and Combination Subfloor
Underlayment — 98

■ **R505.1.1**
Applicability Limits for Steel
Floor Framing — 101

■ **R505.1.3 and R505.3.2**
Floor Trusses (R505.1.3) and Allowable
Joist Spans (R505.3.2) — 103

■ **R506.2.4**
Reinforcement Support — 105

■ **Table R602.3(1)**
Fastener Schedule for Structural Members — 106

■ **Table R602.3(2)**
Alternate Attachments — 110

■ **R602.3.2**
Top Plate — 112

■ **R602.6.1**
Drilling and Notching of Top Plate — 114

■ **R602.10.6.1 and Table R602.10.6**
Alternate Braced Wall Panels — 116

■ **R602.10.6.2**
Alternate Bracing Wall Panel Adjacent   119
to a Door or Window Opening

■ **R602.10.11, R602.10.11.1 through**
R602.10.11.5
Bracing in Seismic Design   122
Categories D_0, D_1, and D_2

■ **R602.11.1**
Wall Anchorage   125

■ **R603.3.2**
Load-Bearing Walls   127

■ **R606.3 and R606.4**
Corbelled Masonry   129

■ **Table R611.2**
Requirements for ICF Walls   131

■ **R613.1**
General Window Installation Instructions   133

■ **R613.2**
Window Sills   134

■ **R613.7**
Wind-borne Debris Protection   136

■ **R613.7.1**
Fenestration Testing and Labeling   138

■ **R613.9.1**
Mullions   139

■ **R702.3.7**
Horizontal Gypsum Board   141
Diaphragm Ceilings

■ **R702.4.2**
Cement, Fiber-Cement, and Glass   143
Mat Gypsum Backers

■ **R703.1**
General Draining Exterior   145
Wall Assemblies

■ **R703.2 and Table R703.4**
Water-Resistive Barrier (R703.2)   147
and Weather-Resistant Siding Attachment
and Minimum Thickness (R703.4)

■ **R703.6.3**
Water-Resistive Barriers   153

■ **R703.7**
Stone and Masonry Veneer, General   155

■ **R703.8**
Flashing   160

■ **R802.1.5**
Structural Log Members   162

■ **R802.3.1 and Table R802.5.1(9)**
Ceiling Joist and Rafter Connections   163

■ **Table R802.5.1(1) through Table R802.5.1(8)**
Rafter Spans for Common Lumber Species   166

■ **R802.10.2.1**
Applicability Limits Wood Truss Design   169

■ **R806.4**
Conditioned Attic Assemblies   171

■ **R903.5, R903.5.1, R903.5.2, and**
Figure R903.5
Hail Exposure (R903.5), Moderate   173
Hail Exposure (R903.5.1), Severe Hail
Exposure (R903.5.2), Hail Exposure Map
(Figure R903.5)

■ **R905.2.6**
Attachment of Asphalt Shingles   175

■ **R905.2.7.1**
Ice Barrier   177

■ **Tables R905.10.3(1) and R905.10.3(2)**
Metal Roof Coverings Standards   179

■ **R905.12.2, R905.13.2, and R905.15.2**
Material Standards   182

■ **R907.3**
Recovering Versus Replacement   184

■ **R1003.15**
Flue Area (Masonry Fireplace)   186

PART 4
Energy Conservation (IRC Chapter 11)   189

■ **Chapter 11**
Energy Efficiency   190

■ **N1101.2**
Compliance   191

■ **N1101.4 and N1101.4.1**
Building Thermal Envelope Insulation   193
(N1101.4) and Blown or Sprayed Roof/
Ceiling Insulation (N1101.4.1)

■ **N1101.8**
Certificate   195

■ **N1102.1 and N1102.1.3**
Insulation and Fenestration Criteria   197
(N1102.1) and Total UA Alternative
(N1102.1.3)

■ **N1102.1.2 and N1102.2.3**
U-Factor Alternative (N1102.1.2)   199
and Mass Walls (N1102.2.3)

PART 5
Mechanical (IRC Chapters 12–23)   202

■ **M1305.1**
Appliance Access for Inspection, Service,   204
Repair, and Replacement

■ **M1305.1.3 and M1305.1.4**
Appliances in Attics (M1305.1.3) and   206
Appliances under Floor (M1305.1.4)

■ **M1308.3**
Foundations and Supports   208

■ **M1411.3.1 and M1411.3.1.1**
Auxiliary and Secondary Drain System   209
(M1411.3.1) and Water Level Monitoring
Devices (M1411.3.1.1)

■ **M1411.4**
Auxiliary Drain Pan   211

■ **M1501.1 and M1506.2**
Outdoor Discharge (M1501.1)   213
and Recirculation of Air (M1506.2)

■ **M1502.6**
Duct Length   215

■ **M1601.3.1**
Joints and Seams   217

■ **Tables M2101.1 and M2101.9, and Sections**
M2103.1, M2103.2, and M2104.2
Hydronic Piping Materials   218
(Table M2101.1), Hanger Spacing Intervals
(Table M2101.9), Piping Materials (M2103.1),
Piping Joints (M2103.2), and Piping Joints
for Low-Temperature Piping (M2104.2)

PART 6
Fuel Gas (IRC Chapter 24)   221

■ **G2403**
Definitions: Point of Delivery   222

■ **G2404.3**
Listed and Labeled   223

■ **G2404.10**
Auxiliary Drain Pan   224

■ **G2415.1**
Prohibited Locations for Gas Piping   226

■ **G2415.5 and G2426.7**
Protection against Physical Damage   227

■ **G2415.6**
Piping in Solid Floors   229

■ **G2420.1.1 and Table G2420.1.1**
Valve Approval (G2420.1.1) and Manual   230
Gas Valve Standards (Table G2420.1.1)

■ **G2421.3, G2421.3.1, and G2403**
Venting of Regulators (G2421.3), Vent   231
Piping (G2421.3.1), and Vent Piping:
Breather and Relief (G2403)

■ **G2422.1**
Connecting Appliances   232

PART 7
Plumbing (IRC Chapters 25–32)   234

■ **P2708.1 and P2708.1.1**
Shower Compartment General (P2708.1)   236
and Access (P2708.1.1)

■ **P2708.3**
Shower Control Valves   238

■ **P2713.3**
Bathtub and Whirlpool Bathtub Valves   239

■ **P2903.4**
Thermal Expansion Control   240

■ **P2904.3**
Polyethylene Plastic Piping Installation   242

■ **P2904.4**
Water Service Pipe   244

■ **P2904.5.1**
Under Concrete Slabs   246

■ **P2904.10**
Polypropylene Plastic   248

■ **Tables P3002.1(1), P3002.1(2), P3002.2,**
and P3002.3
Above-Ground Drainage and Vent Pipe   249
(Table P3002.1(1)), Underground Building
Drainage and Vent Pipe (Table P3002.1(2)),
Building Sewer Pipe (Table P3002.2), and
Pipe Fittings (Table P3002.3)

■ **P3003**
Joints and Connections 252

■ **P3102**
Vent Stacks and Stack Vents 254

■ **P3103.1**
Roof Extension 256

■ **P3105.2**
Fixture Drains 257

■ **P3108.1**
Wet Vent Permitted 259

■ **P3108.2 and P3108.3**
Vent Connections (P3108.2) 261
and Size (P3108.3)

■ **P3201.6**
Number of Fixtures per Trap 263

■ **Table P3201.7**
Size of Traps and Trap Arms 265
for Plumbing Fixtures

PART 8
Electrical (IRC Chapters 33–42) 268

■ **E3605.4.4**
Conductor Sizing of Type NM Cable 270

■ **E3606.2**
Panelboard Circuit Identification 271

■ **E3702.2.2, Table E3702.1**
Protection of Cables Parallel 272
to Furring Strips

■ **E3702.3.2**
Protecting Type NM Cable 274
from Physical Damage

■ **E3702.3.3**
Conductors and Cables Exposed 275
to Direct Sunlight

■ **E3801.3**
Small-Appliance Circuit 276
Receptacle Outlets

■ **E3801.4.1, E3801.4.2**
Receptacle Outlets at Wall and Island 278
Counter Spaces

■ **E3801.6**
Bathroom Receptacles 280

■ **E3801.11**
Heating, Air-Conditioning, 281
and Refrigeration Equipment Outlet

■ **E3802.7**
Ground-Fault Circuit-Interrupter 283
Protection at Laundry and Utility Sinks

■ **E3802.12**
Arc-Fault Circuit-Interrupter Protection 285
in Bedrooms

■ **E3806.8.2.1**
Enclosure Support Using Nails 287
and Screws

■ **E3807.2**
Enclosures in Wet Locations 288

■ **E3807.4**
Repairing Drywall at Panelboards 289

■ **E3902.11, E3902.12**
Receptacles in Wet Locations 291

■ **E3903.10**
Luminaires in Bathtub and Shower Areas 292

■ **E4103.3**
Disconnecting Means for Pool 294
Utilization Equipment

■ **E4106.5.1**
Servicing of Wet-Niche Luminaires 295

Index **297**

Preface

Building officials, design professionals, contractors, and others in the field of residential building construction recognize the need for a modern, up-to-date residential code addressing the design and installation of building systems through requirements emphasizing performance. The *International Residential Code* (IRC), in the 2006 edition, is intended to meet these needs through model code regulations that safeguard the public health and safety in all communities, large and small. The IRC is kept up to date through the open code-development process of the International Code Council (ICC). The provisions of the 2003 edition, along with those code changes approved through 2005, make up the 2006 edition.

The ICC, publisher of the family of International Codes®, was established in 1994 as a nonprofit organization dedicated to developing, maintaining, and supporting a single set of comprehensive and coordinated national model construction codes. Its mission is to provide the highest quality codes, standards, products, and services for all concerned with the safety and performance of the built environment.

The IRC is 1 of 14 International Codes® published by the ICC. This comprehensive residential code establishes minimum regulations for residential building systems by means of prescriptive and performance-related provisions. It is founded on broad-based principles that make possible the use of new materials and new building designs. The IRC is a comprehensive code containing provisions for building, energy, mechanical, fuel gas, plumbing, and electrical systems. The IRC is available for adoption and use by jurisdictions internationally. Its use within a governmental jurisdiction is intended to be accomplished through adoption by reference, in accordance with proceedings establishing the jurisdiction's laws.

The purpose of *Significant Changes to the International Residential Code 2006* is to familiarize building officials, fire officials, plans examiners, inspectors, design professionals, contractors, and others in the building construction industry with many of the important changes in the 2006 IRC. This publication is designed to assist those code users in identifying the specific code changes that have occurred and, more important, understanding the reasons behind the changes. It is also a valuable resource for jurisdictions in their code-adoption process.

Only a portion of the total number of code changes to the IRC are discussed in this book. The changes selected were identified for a number of reasons, including their frequency of application, special significance, or change in application. However, the importance of those changes not included is not to be diminished. Further information on all code changes can be found in the *Code Changes Resource Collection,* published by the ICC. The resource collection provides the published documentation for each successful code change contained in the 2006 IRC since the 2003 edition.

The Significant Changes to the IRC 2006 is organized into eight parts, each representing a distinct grouping of code topics. It is arranged to follow the general layout of the IRC, including code sections and section number format. Throughout the book, each change is accompanied by either a photograph or an illustration to assist and enhance the reader's understanding of the specific change. A summary and a discussion of the significance of the changes are also provided. The code change itself is presented in a format similar to the style utilized for code-change proposals. Deleted language is shown with a strike-through, whereas new code text is indicated by underlining.

The table of contents, in addition to providing guidance in use of this publication, also provides an efficient approach to the identification of those significant code changes that occur in the 2006 IRC.

As with any code-change text, *Significant Changes to the International Residential Code 2006* is best used as a study companion to the 2006 IRC. Because only a limited discussion of each change is provided, the code itself should always be referenced in order to gain a more comprehensive understanding of the code change and its application.

The commentary and opinions set forth in this text are those of the authors and do not necessarily represent the official position of the ICC. In addition, they may not represent the views of any enforcing agency, as such agencies have the sole authority to render interpretations of the IRC. In many cases, the explanatory material is derived from the reasoning voiced by the code-change proponent.

Acknowledgments

The assistance of the City of Austin, Texas, Building Development Services—specifically, staff members Jim Dillinger, Carl Muse, and Dan Garcia—in accommodating residential field trips is greatly appreciated. Thanks also to ICC staff members Gregg Gress, John Henry, Jay Woodward, and Bob Guenther for their usual expert counsel.

About the Author

Hamid A. Naderi, PE, CBO
International Code Council
Principal Staff Engineer

Hamid A. Naderi is a Principal Staff Engineer with the International Code Council (ICC), where he is responsible for the research and development of technical resources and for managing the development of multiple technical projects by expert authors. In addition, Mr. Naderi develops and presents building code, residential code, and existing building code technical seminars nationally and served as the secretariat to the ICC drafting committee for development of the *International Existing Building Code* (IEBC). Prior to joining the ICC through the International Conference of Building Officials (ICBO) in 1995, he spent 12 years with the City of Carrollton, Texas, serving in various technical and management capacities in plan review and field inspection divisions. He also has a broad range of experience in the areas of geotechnical investigation, construction materials testing, and special inspections through his prior employment in the field of geotechnical engineering and construction materials testing. A graduate of the University of Texas in Civil Engineering and a licensed professional engineer, Mr. Naderi obtained his Master of Science Degree with an emphasis in structural engineering and has over 25 years of experience in various areas of building codes administration and training. He is a certified building official and holds three other certifications with the ICC in the areas of building inspection, plumbing, and plan examination.

About the Contributor

Douglas W. Thornburg, AIA, CBO
International Code Council
Director of Product Development

Douglas W. Thornburg is the Director of Product Development for the International Code Council (ICC), where he provides leadership in technical development and positioning of support products for the council. In addition, Mr. Thornburg develops and reviews technical products, reference books, and resource materials relating to the International Codes® and their supporting documents. Prior to employment with the ICC in 2004, he spent 9 years as a code consultant and educator on building codes. Formerly Vice-President/Education for the International Conference of Building Officials (ICBO), Mr.

Thornburg continues to present building code seminars nationally and has developed numerous educational texts and resource materials. He was presented with ICBO's prestigious A. J. (Jack) Lund Award in 1996, in recognition of his outstanding contributions to education and training. A graduate of Kansas State University and a registered architect, Mr. Thornburg has over 25 years of experience in building code training and administration, including 10 years with the ICBO and 5 years with the City of Wichita, Kansas. He is certified as a building official, building inspector, and plans examiner, as well as in seven other code enforcement categories.

SIGNIFICANT CHANGES TO THE
INTERNATIONAL
RESIDENTIAL CODE

2006 EDITION

PART 1

Administration
Chapter 1

- **Chapter 1** Administration

The administration part of the *International Residential Code (IRC)* covers the general scope, purpose, applicability, and other administrative issues related to the regulation of residential buildings by building safety departments. The administrative provisions are the guiding light for the entire code and make clear the responsibilities and duties of various parties involved in residential construction and the applicability of technical provisions within a legal, regulatory, and code enforcement arena.

Section R101.2 establishes the criteria for buildings that are regulated under the IRC. Buildings beyond the scope of Section R101.2 are regulated by the *International Building Code (IBC)*. The IRC governs detached one- and two-family dwellings and townhouses that are not more than three stories in height and have their own separate means of egress. Buildings accessory to such buildings are also regulated by the IRC.

The remaining topics in Part 1 deal with subjects such as the duties and powers of the building official, permits, submittal of construction documents, inspections, the board of appeals, and other such administrative issues. ■

R101.2
Scope

R108.2
Schedule of Permit Fees

R110.1
Use and Occupancy

R101.2

Scope

CHANGE TYPE. Clarification

CHANGE SUMMARY. The revised language allows three-story one- and two-family dwellings and townhouses with basements to be regulated under the *International Residential Code* (IRC). Additionally, the exception that mandated building officials to allow the use of the International Existing Building Code (IEBC) as an alternative to other IRC provisions governing existing buildings has been deleted.

2006 CODE: R101.2 Scope. The provisions of the *International Residential Code for One- and Two-Family Dwellings* shall apply to the construction, alteration, movement, enlargement, replacement, repair, equipment, use and occupancy, location, removal, and demolition of detached one- and two-family dwellings and ~~multiple single-family dwellings (~~townhouses~~)~~ not more than three stories above-grade in height with a separate means of egress and their accessory structures.

> **Exception:** ~~Existing buildings undergoing repair, alteration or additions, and change of occupancy shall be permitted to comply with the *International Existing Building Code*.~~

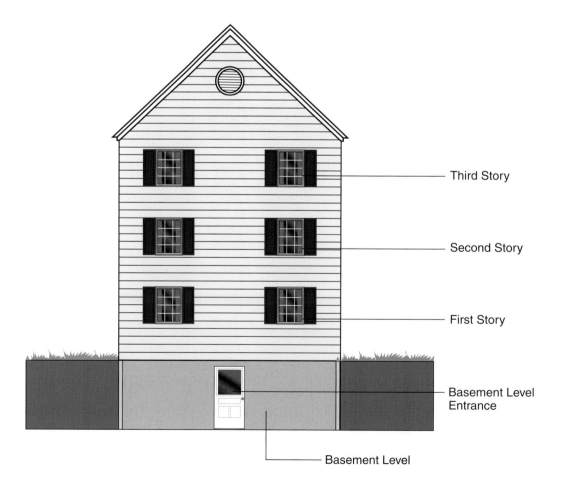

— Third Story

— Second Story

— First Story

— Basement Level Entrance

— Basement Level

CHANGE SIGNIFICANCE.   The term *story* is defined in the IRC (discussed later), and on the basis of that definition, every basement is considered a story. In the 2003 code, for the purposes of the Scoping Section 101.2, a two-story building with one basement would be considered a three-story building, and a three-story building with a basement would be considered a four-story building. It is clear, then, that under the 2003 IRC a three-story building with basement would be considered beyond the scope of the IRC and would have to be regulated by the *International Building Code* (IBC). It was never the intent of the IRC to exclude three-story buildings with a basement from being regulated under the IRC, and therefore the 2006 code change inserts the words "above-grade" so that the three-story upper limit for the IRC applies to three stories above grade. *Story above grade* is also defined in the IRC. The deletion of the phrase "multiple single family dwellings" has no technical implications; this extra phrase is not regularly used in the code.

The following are definitions for story and story above grade. There were no changes related to these definitions between the 2003 and 2006 codes.

STORY. That portion of a building included between the upper surface of a floor and the upper surface of the floor or roof next above.

STORY ABOVE GRADE. Any story having its finished floor surface entirely above grade, except that a basement shall be considered as a story above grade where the finished surface of the floor above the basement is:

1. More than 6 feet (1829 mm) above grade plane;
2. More than 6 feet (1829 mm) above the finished ground level for more than 50% of the total building perimeter; or
3. More than 12 feet (3658 mm) above the finished ground level at any point.

An additional change to this section is the deletion of the exception. Prior to the publication of the 2003 International Existing Building Code (IEBC), proposed changes to other I-Codes attempted to delete any I-Code provision related to the regulation of existing buildings and merely make reference to the IEBC. At that time, many had concerns about various concepts and technical parts of the IEBC and consequently not all of the proposed code changes for this coordination effort were successful. Ultimately, most of the other 2003 I-Codes were coordinated with the 2003 IEBC by an exception in the scoping provisions mandating that the building officials must allow the use of the IEBC for the regulation of existing buildings. During the subsequent code change cycles between 2003 and 2006, in the final action hearings of 2005 held in Detroit, the membership finally decided to remove this exception, and allow each jurisdiction to decide on their own how to or not to incorporate the use of the IEBC. In the IRC, this deletion has also taken place in Section G2401.1, the scoping provisions of fuel gas systems.

R108.2

Schedule of Permit Fees

CHANGE TYPE. Addition

CHANGE SUMMARY. There is now a new Appendix L, which provides a table of permit fees.

2006 CODE: R108.2 Schedule of Permit Fees. On buildings, structures, electrical, gas, mechanical, and plumbing systems or alterations requiring a permit, a fee for each permit shall be paid as required, in accordance with the schedule as established by the applicable governing authority.

CHANGE SIGNIFICANCE. Even though the text of Section R108.2 has not changed, there is now a new Appendix L related to schedule of permit fees. Governmental jurisdictions charge for permit fees to recover the cost of services provided for plan review, inspections, and other activities related to the permit. Many jurisdictions that are adopting a building code for the first time or jurisdictions that need some guidance will now have a resource available contained within Appendix L. Below is the full text of Appendix L.

APPENDIX L Permit Fees

Total Valuation	Fee
$1 to $500	$24
$501 to $2000	$24 for the first $500; plus $3 for each additional $100 or fraction thereof, to and including $2000
$2001 to $40,000	$69 for the first $2000; plus $11 for each additional $1000 or fraction thereof, to and including $40,000
$40,001 to $100,000	$487 for the first $40,000; plus $9 for each additional $1000 or fraction thereof, to and including $100,000
$100,001 to $500,000	$1027 for the first $100,000; plus $7 for each additional $1000 or fraction thereof, to and including $500,000
$500,001 to $1,000,000	$3827 for the first $500,000; plus $5 for each additional $1000 or fraction thereof, to and including $1,000,000
$1,000,001 to $5,000,000	$6327 for the first $1,000,000; plus $3 for each additional $1000 or fraction thereof, to and including $5,000,000
$5,000,001 and over	$18,327 for the first $5,000,000; plus $1 for each additional $1000 or fraction thereof

CHANGE TYPE. Addition

CHANGE SUMMARY. A new exception is added to exempt accessory buildings or structures from the certificate of occupancy requirements.

2006 CODE: R110.1 Use and Occupancy. No building or structure shall be used or occupied, and no change in the existing occupancy classification of a building or structure or portion thereof shall be made until the building official has issued a certificate of occupancy therefore as provided herein. Issuance of a certificate of occupancy shall not be construed as an approval of a violation of the provisions of this code or of other ordinances of the jurisdiction. Certificates presuming to give authority to violate or cancel the provisions of this code or other ordinances of the jurisdiction shall not be valid.

R110.1 continues

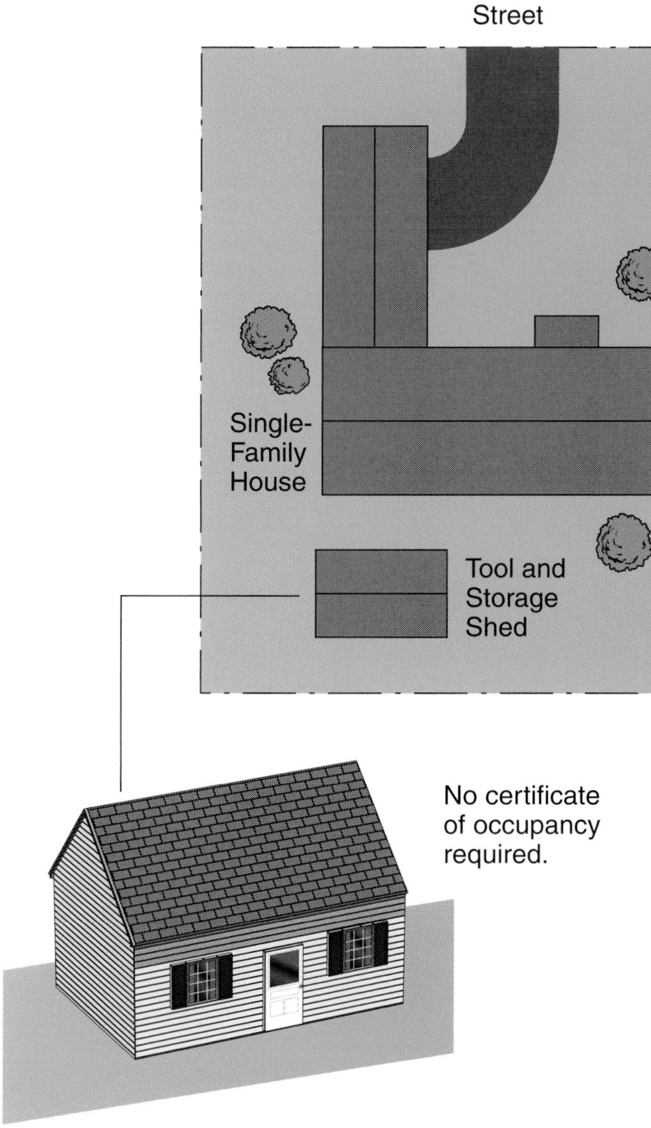

R110.1

Use and Occupancy

Street

Single-Family House

Tool and Storage Shed

No certificate of occupancy required.

R110.1 continued

Exceptions:

1. Certificates of occupancy are not required for work exempt from permits under Section R105.2.

2. <u>Accessory buildings or structures</u>

CHANGE SIGNIFICANCE. Some of the legacy model building codes did not require a certificate of occupancy for accessory structures. These nonhabitable structures are frequently put to use by homeowners with no apparent risk. Lenders, builders, and homeowners never inquire about the need for a certificate of occupancy for such accessory structures. The 2003 code exempts from the certificate of occupancy requirement only those buildings or structures that do not require a permit. In contrast, the 2006 code exempts all such accessory buildings or structures from the certificate of occupancy requirement, whether a permit is required or not. For example, under the 2003 code a detached storage building that is two stories or that exceeds 200 square feet is required to have a certificate of occupancy. For the same scenario, the 2006 code does not require a certificate of occupancy. In accordance with the new definition for accessory structures in the 2006 code, another example might be a detached garage of 1500 square feet, where a permit is clearly required, but a certificate of occupancy is not.

PART 2

Definitions

Chapter 2

- **Chapter 2** Definitions

The definitions contained within the IRC are intended to reflect the special meaning of such terms within the scope of the IRC. As terms can often have multiple meanings within their ordinary day-to-day use or within the various disciplines of the construction industry, it is imperative that their meaning within the context of the IRC be given.

Section R201.3 requires that where terms are not defined in the IRC, the meaning for such a term be taken from the other codes published by the International Code Council (ICC), such as the IBC or others in the family of codes developed by the ICC (I-Codes). Section R201.4 requires that a term not defined anywhere within the family of I-Codes must have the ordinarily accepted meaning such as that which the context implies.

The IRC definitions are contained within Chapter 2; however, terms specifically related to fuel gas and electrical systems are defined in specific chapters, Chapter 24 for fuel gas and Chapter 34 for electrical. Terms defined in Chapters 24 and 34 are not necessarily repeated within Chapter 2. ■

R202 DEFINITIONS
Accessory Structure

R202 DEFINITIONS
Approved

R202 DEFINITIONS
Exterior Wall

R202 DEFINITIONS
Fire-Separation Distance

R202
Definitions
Accessory Structure

CHANGE TYPE.   Modification

CHANGE SUMMARY.   The unnecessary IRC main scoping provisions have been deleted, and specific limitations concerning size and number of stories have been added for accessory structures.

2006 CODE: Accessory Structure.   ~~In one- and two-family dwellings not more than three stories high with separate means of egress, a building,~~ A structue not greater than 3000 square feet (279 m^2) in floor area, and not over two stories in height, the use of which is <u>customarily accessory to and</u> incidental to that of the ~~main building~~ <u>dwelling(s)</u> and which is located on the same lot.

CHANGE SIGNIFICANCE.   The 2003 code language unnecessarily repeats the charging and scoping language of Section 101.2, and therefore this unnecessary language has been deleted. This unnecessary language in the 2003 code is not even repeated in its entirety because of the omission of townhouses, and therefore the definition for accessory structures was not applicable to townhouses. By deleting this language and adding specific criteria, the 2006 code provides a comprehensive definition of accessory structures applicable to one- and two-family dwellings as well as townhouses that fall under the jurisdiction of the IRC. The new definition places a size limitation of 3000

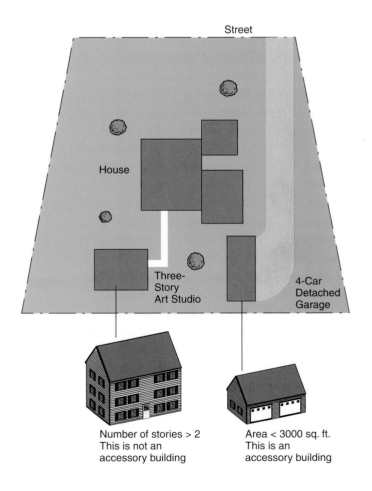

Number of stories > 2
This is not an
accessory building

Area < 3000 sq. ft.
This is an
accessory building

square feet (279 m^2) and a story limitation of two stories and emphasizes that these buildings are subordinate to the dwellings by insertion of the word "accessory." Buildings larger than the limitations provided in the definition will no longer be considered accessory to dwellings, and their use must be evaluated for deciding whether they will fall under the IRC scoping authority or be regulated by the IBC.

R202 Definitions

Approved

CHANGE TYPE.   Modification

CHANGE SUMMARY.   The definition of "approved" has been changed to "acceptable to the building official," with the deletion of other criteria such as conducting investigations or tests.

2006 CODE: Approved. ~~Approved refers to approval by the building official as the result of investigation and tests conducted by him or her, or by reason of accepted principles or tests by nationally recognized organizations.~~ <u>Acceptable to the building official.</u>

CHANGE SIGNIFICANCE.   This change makes the definition of "approved" in the IRC consistent with the definition in the IBC. The concern with the 2003 code definition is the implication that the building official is obligated to conduct tests. This is inconsistent with most of the previous model codes, including the IBC, which defines "approved" as "acceptable to the building official." The new 2006 language does not preclude the building official from requiring tests or any other needed investigations to approve a material, design, or construction method. Tests can be required under IRC Section 104.11.1 and need not be reiterated within the definition.

Acceptable to the Building Official

Building Official:
"The officer or other designated authority charged with the administration and enforcement of this code."

R202
Definitions
Exterior Wall

CHANGE TYPE.   Clarification

CHANGE SUMMARY.   The definition of "exterior wall" has been revised to include all walls that enclose the building and not just walls that enclose conditioned space.

2006 CODE: Exterior Wall.   An above-grade wall ~~enclosing conditioned space~~ that defines the exterior boundaries of a building. Includes between floor spandrels, peripheral edges of floors, roof and basement knee walls, dormer walls, gable end walls, walls enclosing a mansard roof, and basement walls with an average below-grade wall area that is less than 50% of the total opaque and non-opaque area of that enclosing side.

CHANGE SIGNIFICANCE.   Exterior walls and their openings, projections, and penetrations are regulated for fire-resistance rating or fire protection on the basis of the concept of fire-separation distance. These provisions are found in Section R302 and require fire-resistance-rated exterior walls, require protected projections, or prohibit openings in exterior walls that might be constructed closer than a certain distance to lot lines. The 2003 IRC definition for an exterior wall is directly related to "conditioned space." As such, walls enclos-

R202 Definitions continues

Street

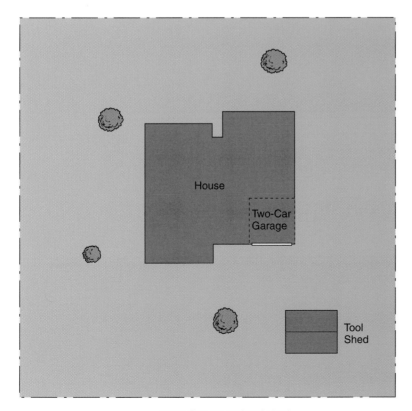

Note: Garage and tool shed
are not conditioned spaces

R202 Definitions continued   ing most garages, tool sheds, or other attached or detached accessory structures would not be considered exterior walls and therefore considered by some to be not regulated at all by Section R302. A review of Section R302 reveals, however, that there are several exceptions specifically addressing garages and other accessory structures. To clarify that Section 302 and all other sections dealing with exterior walls were intended to apply to all exterior walls and not just those exterior walls that enclose conditioned space, the phrase "enclosing conditioned space" was deleted and the phrase "that defines the exterior boundaries of a building" was added.

R202 Definitions
Fire-Separation Distance

CHANGE TYPE. Modification

CHANGE SUMMARY. This modification revises the definition of the term *fire-separation distance* such that the measurement will be taken at right angles to the face of the building exterior wall rather than measuring at right angles to the lot line.

2006 CODE: Fire Separation Distance. The distance measured from the building face to <u>one of the following</u>:

1. To the closest interior lot line; or
2. To the centerline of a street, an alley, or public way; or
3. To an imaginary line between two buildings on the ~~property~~ <u>lot.</u>

The distance shall be measured at right angles from the ~~lot line~~ <u>face of the wall</u>.

CHANGE SIGNIFICANCE. Fire-separation distance is the distance between the face of an exterior wall of a building and the nearest interior lot line, the centerline of a street, alley, or public way, or the imaginary and assumed line between two buildings on the same lot. Fire-separation distance is the trigger mechanism for the determination of the required fire-resistance rating of exterior walls, opening protection, and other issues related to protection of the building envelope from the radiation heat of adjacent buildings. In the 2000 editions of the IBC and IRC, the fire-separation distance is measured at right an-

R202 Definitions continues

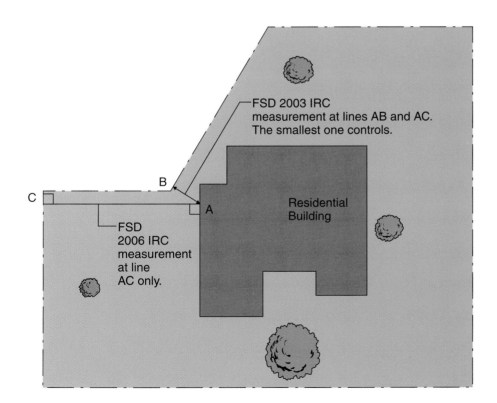

FSD 2003 IRC measurement at lines AB and AC. The smallest one controls.

B
C
A
Residential Building
FSD 2006 IRC measurement at line AC only.

R202 Definitions continued

gles to the lot line rather than at right angles to the face of the wall. In the code change cycles leading to the 2003 editions of the I-Codes, this definition was changed in the 2003 IBC to indicate measurement at right angles to the face of the wall, but the same definition in the 2003 IRC remained unchanged. As such, the measurement methods for the term *fire-separation distance* in the 2003 IBC and the 2003 IRC are not consistent. The new modification basically makes the definition in the 2006 IRC consistent with that of the IBC. Some of the original basis for the change in the 2003 IBC was based on reasons such as: measuring perpendicular to the lot line sometimes results in two measurements for the same point on the exterior wall, and in measurements of the distance between the face of the wall and the lot line the starting point is the face of the wall; thus, such measurements should be taken at right angles to the wall rather than the lot line. This change in the methodology of measurement will not cause a change in the code requirements in most cases. However, in certain cases, it does change the technical application and code requirement. See the accompanying figures for examples, where methods of measurement based on the 2003 IRC and 2006 IRC are both shown. While it is practical to make such measurements at a point on the wall that is closest to the lot line, in many cases it is necessary to make such measurements at other points along the wall where windows or doors are located.

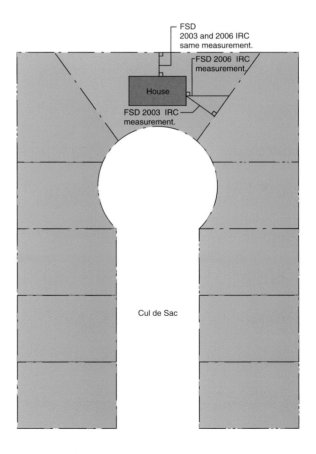

PART **3**

Building Planning and Construction

Chapters 3 through 10

- **Chapter 3**   Building Planning
- **Chapter 4**   Foundations
- **Chapter 5**   Floors
- **Chapter 6**   Wall Construction
- **Chapter 7**   Wall Covering
- **Chapter 8**   Roof-Ceiling Construction
- **Chapter 9**   Roof Assemblies
- **Chapter 10** Chimneys and Fireplaces

Part 3 deals with the overall issues of building planning, design, site location, fire safety, egress, structural system, and other such related issues. Chapter 3 includes the bulk of nonstructural issues such as the location on the lot, light and ventilation, emergency escape and rescue, smoke alarms, and many other issues. Section R308 covers glazing materials and the locations where safety glazing is required, and Sections R311, R312, and R313 address means of egress, guards, and smoke alarms. In addition to such issues, Chapter 3 provides the overall structural loads for the design of residential buildings. Section R301, Design Criteria, addresses all structural loading issues of live loads, dead loads, and other environmental loads such as wind, seismic, and snow.

Other chapters within Part III address the prescriptive as well as the performance criteria for building foundations, floor construction, wall construction, wall coverings, roof construction, roof assemblies, chimneys, and fireplaces. ■

R301.2.1.1

Design Criteria

R301.2.1.2 AND TABLE R301.2.1.2

Protection of Openings

R301.2.2, TABLE R301.2.2.1.1, AND FIGURE R301.2(2)

Seismic Provisions

R301.2.2.2.2

Irregular Buildings

TABLE R301.5

Minimum Uniformly Distributed Live Loads

R302

Exterior Wall and Opening Protection

R303.6.1

Light Activation at Stairways

R305.1

Minimum Height of Sloped Ceilings

R308.3

Glazing Materials Permitted in Hazardous Locations

R308.4

Glazing Adjacent to Stairways and Landings

R309.2

Separation of Detached Garage from Dwelling

R310.1

Emergency Escape and Rescue Openings

R310.1.4, R310.4

Operation of Emergency Escape
and Rescue Openings

R310.5

Emergency Openings under Decks and Porches

R311.4.3

Landings at Exterior Doors

R311.5.4

Landings at Garage Stairways

R311.6.1

Maximum Slope of Ramps

R312.1

Guards at Elevated Ramps

R313.1

Smoke Alarms and Household Fire Alarm Systems

R313.1.1

Smoke Alarms in Existing Dwellings

R317.1

Fire Separation of Two-Family Dwellings

R319.1, R202

Protection of Wood Members against Decay

R319.1.5

Protection of Glued-Laminated Members
against Decay

R320, TABLE R301.2(1)

Protection against Subterranean Termites

R324.1.3.1

Determination of Design Flood Elevations

R401.3

Drainage

R401.4.2

Compressive or Shifting Soil

TABLE R402.2

Minimum Specified Compressive Strength
of Concrete

R403.1.4.2

Seismic Conditions

R403.1.6.1

Foundation Anchorage in Seismic Design
Categories C, D_0, D_1, and D_2

R404.1

Concrete and Masonry Foundation Walls

R404.5

Retaining Walls

R406.1

Concrete and Masonry Foundation Dampproofing

R406.2

Concrete and Masonry Foundation Waterproofing

R408.2, R408.3

Openings for Under-Floor Ventilation (R408.2)
and Unvented Crawl Space (R408.3)

R502.2.1 AND R602.10.8

Framing at Braced Wall Lines (R502.2.1)
and Connections (R602.10.8)

TABLE R502.5(1)

Girder Spans and Header Spans for Exterior Bearing Walls

R502.13

Fireblocking Required

TABLE R503.2.1.1(1)

Allowable Spans and Loads for Wood Structural Panels for Roof and Subfloor Sheathing and Combination Subfloor Underlayment

R505.1.1

Applicability Limits for Steel Floor Framing

R505.1.3 AND R505.3.2

Floor Trusses (R505.1.3) and Allowable Joist Spans (R505.3.2)

R506.2.4

Reinforcement Support

TABLE R602.3(1)

Fastener Schedule for Structural Members

TABLE R602.3(2)

Alternate Attachments

R602.3.2

Top Plate

R602.6.1

Drilling and Notching of Top Plate

R602.10.6.1 AND TABLE R602.10.6

Alternate Braced Wall Panels

R602.10.6.2

Alternate Bracing Wall Panel Adjacent to a Door or Window Opening

R602.10.11, R602.10.11.1 THROUGH R602.10.11.5

Bracing in Seismic Design Categories D_0, D_1, and D_2

R602.11.1

Wall Anchorage

R603.3.2

Load-Bearing Walls

R606.3 AND R606.4

Corbelled Masonry

TABLE R611.2

Requirements for ICF Walls

R613.1

General Window Installation Instructions

R613.2

Window Sills

R613.7

Wind-borne Debris Protection

R613.7.1

Fenestration Testing and Labeling

R613.9.1

Mullions

R702.3.7

Horizontal Gypsum Board Diaphragm Ceilings

R702.4.2

Cement, Fiber-Cement, and Glass Mat Gypsum Backers

R703.1

General Draining Exterior Wall Assemblies

R703.2 AND TABLE R703.4

Water-Resistive Barrier (R703.2) and Weather-Resistant Siding Attachment and Minimum Thickness (R703.4)

R703.6.3

Water-Resistive Barriers

R703.7

Stone and Masonry Veneer, General

R703.8

Flashing

R802.1.5

Structural Log Members

R802.3.1 AND TABLE R802.5.1(9)

Ceiling Joist and Rafter Connections

TABLE R802.5.1(1) THROUGH TABLE R802.5.1(8)

Rafter Spans for Common Lumber Species

R802.10.2.1

Applicability Limits Wood Truss Design

R806.4

Conditioned Attic Assemblies

R903.5, R903.5.1, R903.5.2, AND FIGURE R903.5

Hail Exposure (R903.5), Moderate Hail Exposure (R903.5.1), Severe Hail Exposure (R903.5.2), Hail Exposure Map (Figure R903.5)

R905.2.6

Attachment of Asphalt Shingles

R905.2.7.1

Ice Barrier

TABLES R905.10.3(1) AND R905.10.3(2)

Metal Roof Coverings Standards

R905.12.2, R905.13.2, AND R905.15.2

Material Standards

R907.3

Recovering Versus Replacement

R1003.15

Flue Area (Masonry Fireplace)

R301.2.1.1
Design Criteria

CHANGE TYPE. Modification

CHANGE SUMMARY. For buildings in areas with basic wind speeds of 100 mph or more in hurricane-prone regions, use of the IRC-prescriptive wind provisions will no longer be allowed.

2006 CODE: R301.2.1.1 Design Criteria. Construction in regions where the basic wind speeds from Figure R301.2(4) equal or exceed 100 mph (45 m/s) in hurricane-prone regions, or 110 miles per hour (49 m/s) elsewhere, shall be designed in accordance with one of the following:

1. American Forest and Paper Association (AF&PA) *Wood Frame Construction Manual for One- and Two-Family Dwellings* (WFCM); or

2. *Southern Building Code Congress International Standard for Hurricane Resistant Residential Construction* (SSTD 10); or

3. *Minimum Design Loads for Buildings and Other Structures* (ASCE-7); or

4. American Iron and Steel Institute (AISI), *Standard for Cold-Formed Steel Framing. Prescriptive Method for One- and Two-family Dwellings* (COFS/PM) with supplement to standard for

	Location	Vmph
— 90	Hawaii	105
— 100	Puerto Rico	145
— 110	Guam	170
— 120	Virgin Islands	145
— 130	American Samoa	125
— 140		
— 150		
— Special Wind Region		

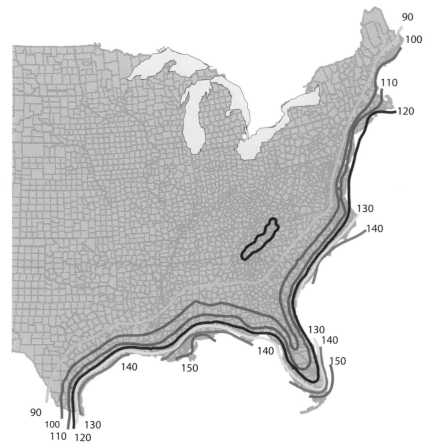

cold-formed steel frames—prescriptive method for one- and two-family dwellings.

5. Concrete construction shall be designed in accordance with the provisions of this code.

CHANGE SIGNIFICANCE.   This change affects any geographical area that has a basic wind speed of 100 mph or more and is located in a hurricane-prone region as defined in Section R202. In the 2003 code the prescriptive provisions can be used in areas where the basic wind speed is up to 110 mph. Beyond this wind speed, the structure must be designed for wind loads on the basis of one of the five standards referenced in this section. The 2006 code, in addition to the 2003 code criteria, adds a new criterion and requires that in hurricane-prone regions with a wind speed of 100 mph or greater, buildings must be designed for wind loads.

Experience has shown that widespread damage is generally associated with hurricane-scale events, whereas thunderstorm-gust fronts are generally localized and short-lived. Therefore, this change to decrease the threshold to 100 mph is specifically aimed only at regions affected by hurricane winds along the Gulf and Atlantic coasts, in addition to Hawaii. Structural analyses demonstrate that there are conditions where toe-nailed connections, wall assembly connections, and wall-bracing requirements are clearly inadequate to resist design wind loads with a reasonable safety margin. The wind uplift concern is most notable for roof slopes less than 6:12 and for longer-span roofs. This concern is true not only for the conventional light-frame wood connections but also for light-gage steel uplift connections. For wall bracing, the issue is primarily a concern with higher roof pitches because the roof structure contributes significantly to the lateral load on the building.

R301.2.1.2 and Table R301.2.1.2

Protection of Openings

CHANGE TYPE. Modification

CHANGE SUMMARY. The title of this section has changed for more clarity and consistency with the IBC. The technical contents of the section have three main revisions. (1) The option of designing buildings as partially enclosed buildings in lieu of protecting windows has been eliminated in windborne debris regions. (2) Other approved methods of Large Missile Tests have been included in addition to the ASTM E1996 and ASTM E1886. (3) The exception has been revised to require the opening protective panels to be connected to the building framing.

2006 CODE: R301.2.1.2 ~~Internal pressure~~ **Protection of Openings.** Windows in buildings located in windborne debris regions shall have glazed openings protected from windborne debris. ~~or the building shall be designed as a partially enclosed building in accordance with the *International Building Code*~~ Glazed opening protection for windborne debris shall meet the requirements of the Large Missile Test of <u>an approved impact resisting standard or</u> ASTM E 1996 and ASTM E 1886 referenced therein.

> **Exception:** Wood structural panels with a minimum thickness of $^7/_{16}$ inch (11.1 mm) and a maximum span of 8 feet (2438

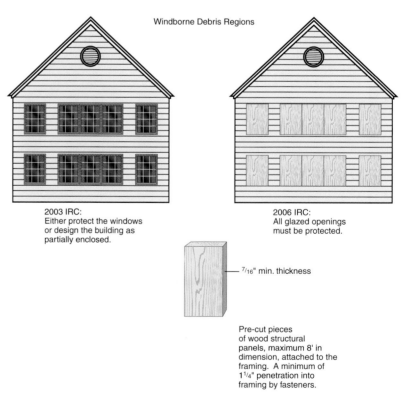

Windborne Debris Regions

2003 IRC:
Either protect the windows
or design the building as
partially enclosed.

2006 IRC:
All glazed openings
must be protected.

$^7/_{16}$" min. thickness

Pre-cut pieces
of wood structural
panels, maximum 8' in
dimension, attached to the
framing. A minimum of
$1^1/_4$" penetration into
framing by fasteners.

Windborne debris region: areas within hurricane-prone
regions within one mile of the costal mean
high water line where the basic wind speed is
110 miles per hour (177 Km/h) or greater; or where the
basic wind speed is equal to or greater than 120 miles per
hour (193 Km/h); or Hawaii.

mm) shall be permitted for opening protection in one- and two-story buildings. Panels shall be precut <u>so that they shall be attached to the framing surrounding the opening containing the product with the glazed opening. Panels shall be secured with the attachment hardware provided.</u> ~~to cover the glazed openings with attachment hardware provided. Attachments shall be provided in accordance with Table R301.2.1.2 or shall be designed to resist the component and cladding loads determined in accordance with the provisions of the *International Building Code.*~~ <u>Attachments shall be designed to resist the components and cladding loads determined in accordance with either Table R301.2(2) or Section 1609.6.5 of the *International Building Code.* Attachment in accordance with Table R301.2.1.2 is permitted for buildings with a mean roof height of 33 feet (10 058 mm) or less where wind speeds do not exceed 130 miles per hour (58 m/s).</u>

TABLE R301.2.1.2 Windborne Debris Protection Fastening Schedule for Wood Structural Panels[a,b,c,d]

Fastener Type	Fastener Spacing (inches)		
	Panel Span ≤ 4 Feet	4 Feet < Panel Span ≤ 6 Feet	6 Feet < Panel Span ≤ 8 Feet
~~2 ⅜~~ No. 6 Screws	16	12	9
~~2 ⅜~~ No. 8 Screws	16	16	12

a. This table is based on 130 mph (3-second gust) and a 33-foot mean roof height.
b. Fasteners shall be installed at opposing ends of the wood structural panel. <u>Fasteners shall be located a minimum of 1″ from the edge of the panel.</u>
c. <u>Fasteners shall be long enough to penetrate through the exterior wall covering and a minimum of 1 ¼″ into wood wall framing and a minimum of 1 ¼″ into concrete block or concrete. Fasteners shall be located a minimum of 2 ½″ from the edge of concrete block or concrete.</u>
~~c~~-d. Where screws are attached to masonry or masonry/stucco, they shall be attached utilizing vibration-resistant anchors having a minimum ultimate withdrawal capacity of 490 pounds.

CHANGE SIGNIFICANCE. This modification eliminates the option of designing a building as partially enclosed in lieu of protecting the exterior windows in windborne debris regions. Design of partially enclosed buildings assumes that the exterior windows will be damaged or broken, which causes internal pressures that must be considered in the design of the Main Wind Force Resisting System in addition to the external wind pressures. Storm damage surveys have long demonstrated the vulnerability of unprotected glazing in high-wind events. The windborne debris environment is severe and includes shingles, tiles, branches, twigs, lawn furniture, wood sheathing, and framing lumber. The most common debris are pieces of failed roof coverings, which originate 15–35 feet above ground in residential areas and are blown into neighboring houses. A high percentage of all glazing failures in residential areas are from windborne debris impacts. The 2006 code requires all glazed openings in windborne debris regions be protected.

R301.2.1.2 and Table R301.2.1.2 continues

*R301.2.1.2 and Table R301.2.1.2
continued*

The 2003 IRC protection of glazed openings is based on test standards of ASTM E 1886 and ASTM E 1996. The purpose of the changes resulting in new language in the 2006 IRC is to make this provision consistent with the current equivalent provision in IBC Section 1609.1.4, "Protection of Openings."

The ASTM E 1886 and ASTM E 1996 are the latest consensus-based testing standards for windborne debris missile impact. However, since 1993, many products have been developed and tested against other impact-resisting standards, including Southern Reference Standard SSTD 12-97, Miami-Dade protocol PA 201-202-203, and Texas Department of Insurance document (TDI) 1-98. All of these standards involve a large missile test where 2x4s of varying lengths are fired at the specimen. These standards vary slightly in the size, speed, and location of the missile, as well as the pressure cycling done after impact. Of these standards, the majority of shutter manufacturers have met either SSTD12-97 or the Miami-Dade PA201-202-203 protocols.

The 2003 IRC, by restricting the language to allow only the ASTM standards, may require shutter manufacturers, whose products have already passed an equivalent test, to retest to the ASTM standard in order to sell products in areas that adopt the IRC.

The proposed revision also accomplishes the following:

■ Requires that the protective panels be secured to the framing that surrounds the opening containing the product with the glazed opening. This requires that the fasteners penetrate into wood wall framing, concrete block, or concrete. The 2003 IRC appears to allow the fasteners that secure the panels to be attached to the exterior opening product itself, which may not consist of materials capable of providing the required withdrawal resistance.

■ Permits the design of fastener attachment in accordance with component and cladding loads specified in the IRC and the IBC.

■ Clarifies the limitations on building height and wind speed for using Table R301.2.1.2. This is consistent with the limitations specified in the IBC.

■ Deletes fastener lengths from Table R301.2.1.2. Since the fasteners may penetrate through exterior coverings of various thickness, a footnote to Table R301.2.1.2 states that the fasteners must be long enough to penetrate a minimum depth into the framing members.

■ Specifies a minimum distance from the edge of the panels for the fasteners.

R301.2.2, Table R301.2.2.1.1, and Figure R301.2(2)

Seismic Provisions

CHANGE TYPE. Addition

CHANGE SUMMARY. This section and related tables and figures have been changed to include an additional seismic design category, D_0, and the seismic design category maps have been updated to reflect the newest information.

2006 CODE: R301.2.2 Seismic Provisions. The seismic provisions of this code shall apply to buildings constructed in Seismic Design Categories C, $\underline{D_0}$, D_1, and D_2, as determined in accordance with this section. Buildings in Seismic Design Category E shall be designed in accordance with the *International Building Code,* except when the Seismic Design Category is reclassified to a lower Seismic Design Category in accordance with Section R301.2.2.1.

> **Exception:** Detached one- and two-family dwellings located in Seismic Design Category C are exempt from the seismic requirements of this code.

The weight and irregularity limitations of Section R301.2.2.2 shall apply to buildings in all Seismic Design Categories regulated by the seismic provisions of this code. Buildings in Seismic Design Category C shall be constructed in accordance with the additional requirements of Sections R301.2.2.3. Buildings in Seismic Design Categories $\underline{D_0}$, D_1, and D_2 shall be constructed in accordance with the additional requirements of Sections R301.2.2.4

TABLE R301.2.2.1.1   Seismic Design Category Determination

Calculated S_{DS}	Seismic Design Category
$S_{DS} \leq 0.17$ g	A
$0.17g < S_{DS} \leq 0.33$ g	B
$0.33g < S_{DS} \leq 0.50$ g	C
$\underline{0.50g < S_{DS} \leq 0.67 \text{ g}}$	$\underline{D_0}$
~~0.50 g~~ $\underline{0.67g} < S_{DS} \leq 0.83$ g	D_1
$0.83g < S_{DS} \leq 1.17$ g	D_2
$1.17g < S_{DS}$	E

CHANGE SIGNIFICANCE. The change to this section creates a new seismic design category (SDC), D_0, by dividing the current SDC D_1 into two divisions. The current SDC D_1, which is associated with Short-Period Response Accelerations in the range of 0.50g and 0.83g, has been thought by some to be too broad a range for certain applications within the scope of the IRC. This addition of the new SDC takes place throughout the code in each and every section that contains the SDC D_1 and as a result expands the current six SDCs to seven. The new ranges for SDCs D_0 and D_1 are now as shown in Table R301.2.2.1.1. The net effect of this change places the same requirements for SDC D_1 on all D_0, which means there is not a net change in the requirements of the code intended by this change alone. However, this sets the framework for additional changes to allow some relief in category D_0 if such

R301.2.2, Table R301.2.2.1.1, and Figure R301.2(2) continues

R301.2.2, Table R301.2.2.1.1, and Figure R301.2(2) continued

relief can be justified with a rational analysis of the forces involved, for cases such as reducing the requirements for brick veneer on conventional construction wood light-frame buildings. The series of Figures R301.2(2) have all been revised to add the new category D_0 and have been updated to reflect improved local information on seismic risk for many regions of the United States and compatible new maps for Puerto Rico.

CHANGE TYPE. Clarification

CHANGE SUMMARY. This clarification expands the prohibition against the use of the code-prescriptive provisions for concrete and wood light-frame construction to all construction materials in certain seismic design categories.

2006 CODE: R301.2.2.2.2 Irregular Buildings. ~~Concrete construction complying with Section R611 or R612 and conventional light-frame construction shall not be used in irregular portions of~~ Prescriptive construction as regulated by this code shall not be used for irregular structures located in Seismic Design Categories C, $\underline{D_0}$, D_1 and D_2. ~~Only such i~~ Irregular portions of structures shall be designed in accordance with accepted engineering practice to the extent such irregular features affect the performance of the ~~conventional framing~~ remaining structural system. When the forces associated with the irregularity are resisted by a structural system designed in accordance with accepted engineering practice, the remainder of the building shall be permitted to be designed using the provisions of this code. A building or portion of a building shall be considered to be irregular when one or more of the following conditions occur:

[There are no changes to items 1 through 7, describing irregular buildings and their exceptions.]

CHANGE SIGNIFICANCE. The 2003 code prohibits the use of prescriptive provisions for concrete and wood light-frame construction when the building is considered irregular and is located in Seismic Design Categories (SDCs) C, D_1, and D_2. The code users have always questioned why the prohibition applied to only concrete and wood light-frame and not other construction materials, since the forces associated with irregularity are the result of geometric configurations and not construction materials used. Many took the position that this section of the code had always been intended to apply to all buildings, not just concrete and wood light-frame construction, even if the code text did not indicate this. The change resulting in the current 2006 code language clarifies that all of the material groups are now governed by this limitation. Irregular buildings in SDCs C, D_0, D_1, and D_2

R301.2.2.2.2 continues

R301.2.2.2.2

Irregular Buildings

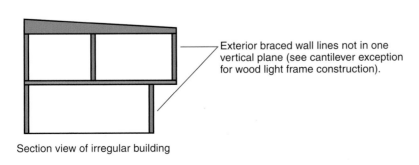

Section view of irregular building

Exterior braced wall lines not in one vertical plane (see cantilever exception for wood light frame construction).

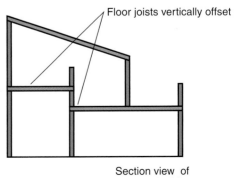

Floor joists vertically offset

Section view of irregular building

R301.2.2.2.2 continued   cannot use the code's prescriptive provisions, regardless of construction material used. The revision further clarifies that once the forces associated with the irregularity of the building are resisted by a designed and engineered structural system, the remainder of the building can still be constructed on the basis of the code-prescriptive provisions; hence, a combination of engineered and conventional construction is permitted in one building.

Table R301.5

Minimum Uniformly Distributed Live Loads

CHANGE TYPE. Modification

CHANGE SUMMARY. The safety factor for glazing in handrail assemblies has been clarified to be 4 for consistency with the IBC, and truss-loading criteria for attics with and without storage have been clarified.

2006 CODE:

TABLE R301.5 Minimum Uniformly Distributed Live Loads (in pounds per square foot)

Use	Live Load
Attics with <u>limited</u> storage[b,g,h]	20
Attics without storage[b]	10
Decks[e]	40
Exterior Balconies	60
Fire Escapes	40
Guardrails and handrails[d]	200[i]
<u>Guardrails in-fill components</u>[f]	<u>50</u>[i]
Passenger vehicle garages[a]	50[a]
Rooms other than sleeping rooms	40
Sleeping rooms	30
Stairs	40[c]

For SI: 1 pound per square foot = 0.0479 kN/m^2, 1 square inch = 645mm^2, 1 pound = 4.45 N.

a. Elevated garage floors shall be capable of supporting a 2000-pound load applied over a 20-square-inch area.

b. ~~No storage with roof slope not over 3 units in 12 units.~~ <u>Attics without storage are those where the maximum clear height between joist and rafter is less than 42 inches, or where there are not two or more adjacent trusses with the same web configuration capable of containing a rectangle 42 inches high by 2 feet wide, or greater, located within the plane of the truss. For attics without storage, this live load need not be assumed to act concurrently with any other live load requirements.</u>

c. Individual stair treads shall be designed for the uniformly distributed live load or a 300-pound concentrated load acting over an area of 4 square inches, whichever produces the greater stresses.

d. A single concentrated load applied in any direction at any point along the top.

e. See Section R502.2.1 for decks attached to exterior walls.

f. Guardrail in-fill components (all those except the handrail), balusters, and panel fillers shall be designed to withstand a horizontally applied normal load of 50 pounds on an area equal to 1 square foot. This load need not be assumed to act concurrently with any other live load requirement.

g. <u>For attics with limited storage and constructed with trusses, this live load need be applied only to those portions of the bottom chord of not less than two adjacent trusses with the same web configuration containing a rectangle 42 inches high or greater by 2 feet wide or greater, located within the plane of the truss. The rectangle shall fit between the top of the bottom chord and the bottom of any other truss member, provided that each of the following criteria is met:</u>
    1. <u>The attic area is accessible by a pull-down stairway or framed opening in accordance with Section R807.1; and</u>
    2. <u>The truss shall have a bottom chord pitch less than 2:12.</u>

h. <u>Attic spaces served by a fixed stair shall be designed to support the minimum live load specified for sleeping rooms.</u>

i. <u>Glazing used in handrail assemblies and guards shall be designed with a safety factor of 4. The safety factor shall be applied to each concentrated load applied to the top of the rail, and to the load on the in-fill components. These loads shall be determined independent of one another, and loads are assumed not to occur with any other live load.</u>

Table R301.5 continues

Table R301.5 continued

CHANGE SIGNIFICANCE.   The 2003 IRC requirements for determining when ceiling elements should be designed for storage loads are not clear and have caused much confusion, especially with regard to truss systems. Roof/ceiling trusses, which are quite common in one- and two-family dwellings and other residential construction, do not typically provide the broad, open spaces that conventional rafter construction provides. Some of the legacy model codes addressed ceiling joists and trusses separately, but the 2003 IRC provisions do not.

The 2003 IRC Footnote 'b' states: "No storage with roof slopes not over 3 units in 12 units." Apart from this footnote, there are no other criteria in the code for attic loading, either for ceiling joists or for trusses. Because of the lack of more precise provisions, Table R301.5 and Footnote 'b' in the 2003 IRC are sometimes interpreted as requiring the bottom chords of residential trusses to be designed to support a live load of 20 psf over the entire span of the bottom chord, regardless of web configuration, clear space, bottom chord slope, and access to the space, all of which will determine whether the space can be used for storage.

This 2006 IRC revised table and its new footnotes will benefit both designers and code officials by providing clear criteria for attic loading for both conventional rafter construction and trusses. It contains three essential criteria:

1. a clear space-dimension threshold at which significant storage could occur and a storage load application of 20 psf;

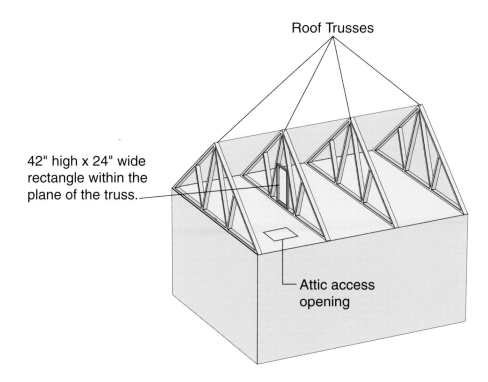

Roof Trusses

42" high x 24" wide rectangle within the plane of the truss.

Attic access opening

Note: Truss bracing members not shown for clarity

2. a minimum live load requirement when storage loads do not apply (because significant storage is precluded by low clearance or web configuration), for occasional access into small spaces (e.g., for maintenance); and

3. a requirement that all attic spaces served by a fixed stair be designed as sleeping room floors.

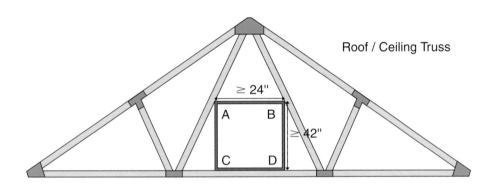

Roof / Ceiling Truss

Limited attic storage
if rectangle ABCD: AC or BD ≥ 42" high and
                            AB or CD ≥ 24" wide
                            and a pull-down stair or
                            attic access opening is provided.

No attic storage:
If any of the above conditions do not exist.

Regular floor design as
a bedroom 30 psf: if a fixed stair serves the attic

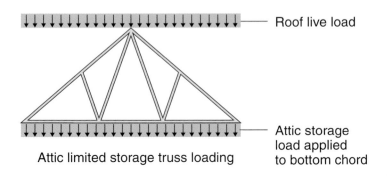

Roof live load

Attic storage
load applied
to bottom chord

Attic limited storage truss loading

R302

Exterior Wall and Opening Protection

CHANGE TYPE. Modification

CHANGE SUMMARY. This code section has been revised for the purpose of both format and technical modification. Exterior walls are now required to have a minimum 1-hour fire-resistance rating where the fire-separation distance is less than 5 feet. In addition, unprotected openings are permitted in limited quantities for fire-separation distances between 3 feet and 5 feet.

2006 CODE: Section ~~302 Location on Lot~~ R302.1 ~~Exterior Walls.~~ ~~Exterior walls with a fire separation distance less than 3 feet (914 mm) shall have not less than a one-hour fire-resistive rating with exposure from both sides. Projections shall not extend to a point closer than 2 feet (610 mm) from the line used to determine the fire separation distance.~~

~~**Exception:** Detached garages accessory to a dwelling located within 2 feet of a lot line may have roof eave projections not exceeding 4 inches.~~

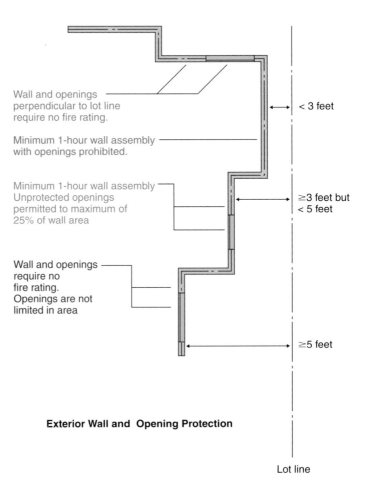

Wall and openings
perpendicular to lot line
require no fire rating.

Minimum 1-hour wall assembly
with openings prohibited.

Minimum 1-hour wall assembly
Unprotected openings
permitted to maximum of
25% of wall area

Wall and openings
require no
fire rating.
Openings are not
limited in area

← < 3 feet

← ≥3 feet but
< 5 feet

← ≥5 feet

Exterior Wall and Opening Protection

Lot line

~~Projections extending into the fire separation distance shall have not less than one hour fire resistive construction on the underside. The above provisions shall not apply to walls which are perpendicular to the line used to determine the fire separation distance.~~

~~**Exception:** Tool and storage sheds, playhouses and similar structures exempted from permits by R105.2 are not required to provide wall protection based on location on the lot. Projections beyond the exterior wall shall not extend over the lot line.~~

~~**R302.2 Openings.** Openings shall not be permitted in the exterior wall of a dwelling or accessory building with a fire separation distance less than 3 feet (914 mm). This distance shall be measured perpendicular to the line used to determine the fire separation distance.~~

~~**Exceptions:**~~
~~1. Openings shall be permitted in walls that are perpendicular to the line used to determine the fire separation distance.~~
~~2. Foundation vents installed in compliance with this code are permitted.~~

~~**R302.3 Penetrations.** Penetrations located in the exterior wall of a dwelling with a fire separation distance less than 3 feet (914 mm) shall be protected in accordance with Section R317.3.~~

~~**Exception:** Penetrations shall be permitted in walls that are perpendicular to the line used to determine the fire separation distance.~~

Section R302 Exterior Wall Location; R302.1 Exterior Walls.

Construction, projections, openings, and penetrations of exterior walls of dwellings and accessory buildings shall comply with Table R302.1. These provisions shall not apply to the walls, projections, openings, or penetrations in walls that are perpendicular to the line used to determine the fire separation distance. Projections beyond the exterior wall shall not extend more than 12 inches (305 mm) into the areas where openings are prohibited.

Exceptions:
1. Detached tool sheds and storage sheds, playhouses, and similar structures exempted from permits are not required to provide wall protection based on location on the lot. Projections beyond the exterior wall shall not extend over the lot line.
2. Detached garages accessory to a dwelling located within 2 feet (610 mm) of a lot line may have roof eave projections not exceeding 4 inches (102 mm).
3. Foundation vents installed in compliance with this code are permitted.

R302 continues

R302 continued

TABLE R302.1   Exterior Walls

Exterior Wall Element		Minimum Fire-Resistance Rating	Minimum Fire Separation Distance
Walls	(Fire-resistance Rated)	1-Hour with Exposure from Both Sides	0 Feet (0 mm)
	(Non Fire-resistance Rated)	0-Hours	5 Feet (1525 mm)
Projections	(Fire-resistance Rated)	1-Hour on the Underside	4 Feet (305 mm)
	(Non Fire-resistance Rated)	0-Hours	5 Feet (1525 mm)
Openings	Not Allowed	N/A	≤ 3 Feet (915 mm)
	25% Maximum of Wall Area	0-Hours	3 Feet (915 mm)
	Unlimited	0-Hours	5 Feet (1525 mm)
Penetrations	All	Comply with Section R317.3	≤ 5 Feet (1525 mm)
		Non Required	5 Feet (1525 mm)

N/A = Not Applicable

CHANGE SIGNIFICANCE.   Previously, a minimum fire-separation distance of 3 feet was necessary to eliminate the requirement for a 1-hour fire-resistance-rated exterior wall. The code change increases the distance to 5 feet, making the requirement consistent with that for Group R-3 occupancies in the IBC. Where an exterior wall has a fire separation distance of less than 5 feet, a minimum 1-hour fire-resistance rating is mandated. Penetrations of the exterior wall are also required to be protected within the 5-foot fire-separation distance.

Projections now cannot be located where the fire-separation distance is less than 4 feet. This limitation previously only applied if the fire-separation distance was less than 2 feet. Application of the projection limitation has remained consistent, allowing any projection to extend only 12 inches into the area where the exterior wall is regulated for fire resistance.

Openings in the exterior wall remain prohibited where the fire-separation distance is less than 3 feet. A new provision allows for a limited amount of unprotected openings where the fire-separation distance is between 3 feet and 5 feet. The total surface area of such openings is limited to 25 percent of the total wall area. There are no fire protection requirements for openings once the fire separation distance reaches 5 feet.

In addition to the technical modifications, the format of Section 302 has been totally revised. A new table has been created to better identify the requirements. The criteria for exterior walls, projections, openings, and penetrations are individually addressed to clarify the requirements for each building element.

CHANGE TYPE. Modification

CHANGE SUMMARY. The requirement for a wall switch at each floor level which controls the lighting outlet illuminating an interior stairway is now mandated only where the stairway has a minimum of six risers.

2006 CODE: R303.6.1 Light Activation. ~~The control for activation of the required interior stairway lighting shall be accessible at the top and bottom of each stairway without traversing any steps.~~ <u>Where lighting outlets are installed in interior stairways, there shall be a wall switch at each floor level to control the lighting outlet where the stairway has six or more risers.</u> The illumination of exterior stairways shall be controlled from inside the dwelling unit.

> **Exception:** Lights that are continuously illuminated or automatically controlled.

CHANGE SIGNIFICANCE. All interior stairways must be provided with illumination under the provisions of Section R303.6. Previously, the code required an activating control at both the top and bottom of

R303.6.1 continues

R303.6.1
Light Activation at Stairways

R303.6.1 continued   the stairway, accessible to the user without traversing any steps. The code change now limits the requirement for a wall switch at each floor level to only those stairways having six or more risers. The new language is consistent with the electrical provisions of Section E3803.3 addressing lighting outlets. The change in application will occur primarily in stairs between a garage and dwelling unit, as well as stairs in a split-level dwelling condition where there are a limited number of risers.

CHANGE TYPE.   Clarification

CHANGE SUMMARY.   The code text addressing minimum ceiling height in rooms with sloped ceilings has been revised to clarify that only 50% of a room's required floor area must have a minimum 7-foot ceiling height.

2006 CODE: R305.1 Minimum Height.   Habitable rooms, hallways, corridors, bathrooms, toilet rooms, laundry rooms, and basements shall have a ceiling height of not less than 7 feet (2134 mm). The required height shall be measured from the finish floor to the lowest projection from the ceiling.

Exceptions:

1. and 2. (no change to text)

3. ~~Not more than 50 percent of the required floor area of a room or space is permitted to have a sloped ceiling less than 7 feet (2134 mm) in height with no portion of the required floor area less than 5 feet (1524 mm) in height.~~ For rooms with sloped ceilings, at least 50 percent of the required floor area of the room must have a ceiling height of at least 7 feet (2134 mm)

R305.1 continues

R305.1

Minimum Height of Sloped Ceilings

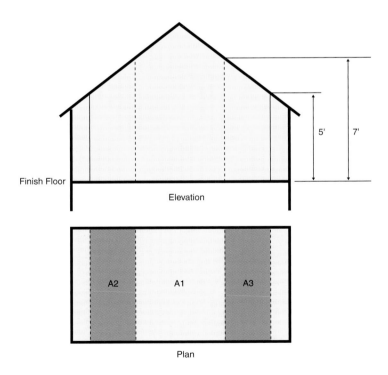

Elevation

Plan

A1 + A2 + A3 ≥ Required room floor area per
Section R304

A1 ≥ 50% of required room floor area
per Section R304

Minimum Height of Sloped Ceilings

R305.1 continued   <u>and no portion of the required floor area may have a ceiling height of less than 5 feet (1524 mm).</u>

4. (no change to text)

CHANGE SIGNIFICANCE.   The code regulates the minimum ceiling height within a dwelling unit at 7 feet, allowing for a reduction in those rooms with a sloped ceiling. Although the entire provision for sloped ceilings has been rewritten, the application is intended to remain unchanged. Where a sloped ceiling occurs within a room, at least one-half of the minimum required floor area must have a minimum ceiling height of 7 feet. For example, if the room is required by Section R304.2 to have at least 70 square feet of floor area, then at least 35 square feet of floor area must be provided with a ceiling height of at least 7 feet. In determining if the room has the minimum 70 square feet initially required, only those portions at least 5 feet in height can be considered. Once these two conditions have been met, the ceiling height of the remaining floor area is unregulated.

R308.3

Glazing Materials Permitted in Hazardous Locations

CHANGE TYPE. Modification

CHANGE SUMMARY. Where glazing occurs in fire doors and fire windows, it must now comply with the test requirements of Consumer Product Safety Commission (CPSC) 16 Code of Federal Regulations (CFR), Part 1201, as the use of polished wired glass tested in accordance with American National Standards Institute (ANSI) Z97.1 is no longer permitted.

2006 CODE: R308.3 Human Impact Loads. Individual glazed areas including glass mirrors in hazardous locations such as those indicated as defined in Section R308.4 shall pass the test requirements of CPSC 16 CFR, Part 1201. Glazing shall comply with CPSC 16 CFR, Part 1201 criteria for Category I or Category II as indicated in Table R308.3.

Exceptions:
1. ~~Polished wired glass for use in fire doors and other fire resistant locations shall comply with ANSI Z97.1.~~
2. Louvered windows and jalousies shall comply with Section R308.2.

CHANGE SIGNIFICANCE. The code has previously permitted the use of polished wired glass complying with ANSI 97.1 in fire doors, fire windows, and view panels in fire-resistance-rated walls. The code change now requires that all glazing, including that considered as wired glass, be tested in accordance with CPSC 16 CFR Part 1201.

Polished wired glass tested under the requirements of ANSI 97.1, although performing well as a fire-protective component, has been

R308.3 continues

R308.3 continued   deemed inadequate for use in locations subject to human impact. Lacking the strength of regular annealed glass, the glazing also presents a potential hazard from exposed wires upon breakage. The use of CPSC 16 CFR Part 1201 as the test standard for all safety glazing materials, including those utilized in fire doors and fire windows, provides for a consistent degree of protection against breakage.

CHANGE TYPE.   Addition

CHANGE SUMMARY.   An additional method of exempting the use of safety glazing adjacent to stairways and landings has been introduced to the code. In addition, new language clarifies that complying glass block panels are also exempt from safety glazing requirements.

2006 CODE: 308.4 Hazardous Locations.   The following shall be considered specific hazardous locations for the purposes of glazing:

1. through 11. (no change to text)

Exception: The following products, materials and uses are exempt from the above hazardous locations:

1. **through 3.** (no change to text)
4. Glazing in Section R308.4, Item 6, in walls perpendicular to the plane of the door in a closed position, <u>other than the wall</u>

R308.4 continues

R308.4

Glazing Adjacent to Stairways and Landings

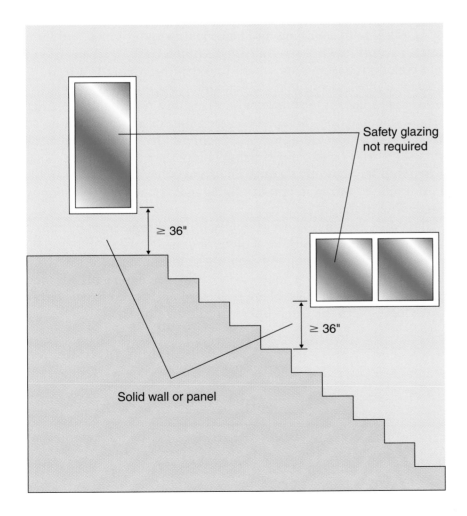

R308.4 continued

toward which the door swings when opened, or where access through the door is to a closet or storage area 3 feet (914 mm) or less in depth. Glazing in these applications shall comply with Section R308.4, Item 7.

5. **through 8.** (no change to text)

9. Safety glazing is Section R308.4, Items 10 and 11, is not required where:

    9.1 The side of a stairway, landing or ramp has a guardrail or handrail, including balasters or in-fill panels, complying with the provisions of Section 1013 and 1607.7 of the *International Building Code;* and

    9.2 The plane of the glass is greater than 18 inches (457 mm) from the railing., or

    9.3 When a solid wall or panel extends from the plane of the adjacent walking surface to 34 inches (863 mm) to 36 inches (914 mm) above the floor and the construction at the top of that wall or panel is capable of withstanding the same horizontal load as the protective bar.

10. Glass block panels complying with Section R610.

CHANGE SIGNIFICANCE. Items 10 and 11 of Section R308.4 regulate glazing adjacent to stairways and landing due to the concern of human impact. Exception 9 has previously exempted such locations from the safety glazing requirements where two conditions (both 9.1 and 9.2) occurred. A new exception has been added that will also exempt those glazed areas adjacent to stairways and landings from being considered as hazardous locations for safety glazing purposes. If the bottom edge of the glazing is located at least 36 inches above the walking surface with a solid wall or panel extending below the glazing, the glazing is not required to be safety glazing. The code change effectively reduces the minimum required height of glazing that is not safety glazing from 60 inches to 36 inches.

Glass unit masonry is occasionally utilized in areas subject to human impact, such as walk-in shower enclosures. Glass block panels are now specifically exempted from the requirements for safety glazing when in compliance with Section R610 (Glass Unit Masonry). Previously, the use of glass block in hazardous locations was only directly addressed in the *International Building Code.*

CHANGE TYPE. Addition

CHANGE SUMMARY. The code now provides for a fire-resistive separation between a dwelling and a detached garage where the structures are located within 3 feet of each other. A new section R309.1.2 titled "Other Penetrations" has also been introduced to require the annular space in the garage/dwelling penetrations be protected with approved materials.

2006 CODE: R309.2 Separation Required. The garage shall be separated from the residence and its attic area by not less than ½-inch (12.7 mm) gypsum board applied to the garage side. Garages beneath habitable rooms shall be separated from all habitable rooms above by not less than ⅝-inch (15.9 mm) type X gypsum board or equivalent. Where the separation is a floor-ceiling assembly, the structure supporting the separation shall also be protected by not less than ½-inch (12.7 mm) gypsum board or equivalent. Garages located less than 3 feet (914 mm) from a dwelling unit on the same lot shall be protected with not less than ½-inch (12.7 mm) gypsum board applied to the interior side of exterior walls that are within this area. Openings in these walls shall be regulated by Section R309.1. This provision does not apply to garage walls that are perpendicular to the adjacent dwelling unit wall.

R309.2 continues

R309.2

Separation of Detached Garage from Dwelling

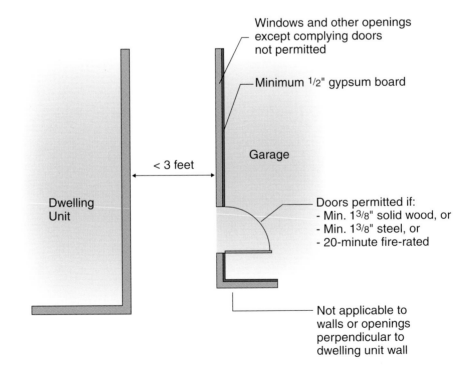

Separation of Detached Garage from Dwelling

R309.2 continued **CHANGE SIGNIFICANCE.** The code has traditionally required some degree of fire-resistive separation between a dwelling and an attached garage, currently mandating a minimum of ½-inch gypsum board on the common wall. However, there were no previous requirements where the garage was not attached directly to the dwelling. This new language now establishes a means of protecting a dwelling from a garage fire in a detached garage where the garage is located within 3 feet of the dwelling. The degree of protection is consistent with that required for attached garages, with a minimum of ½-inch gypsum board applied to the interior of the garage wall.

There is no specific measurement method established for determining which portions of the garage wall must be protected. The code does indicate, however, that the gypsum board application is not required on any garage wall located at a right angle to the wall of the dwelling. It would seem logical that the 3-foot measurement be taken at right angles from the garage wall, rather than from the dwelling, as the provision is based solely upon a fire spreading from the garage to the dwelling.

There is also no provision in the 2003 IRC regulating any door or other opening placed in the wall of either the detached garage or the dwelling. It is expected that the level of protection be similar to that required for an attached garage/dwelling separation, thus in the 2006 IRC limiting any openings to doors complying with Section R309.1, when detached garage is located less than 3 feet to the dwelling. In addition, as a result of the membership approval of the public comment for code change RB66-04105, a new section R309.1.2 has been introduced in the code with language similar to fireblocking provisions of item 4 of R602.8. This is intended to protect the annular space around the penetrating items of the garage/dwelling separation for additional safety.

R310.1
Emergency Escape and Rescue Openings

CHANGE TYPE.   Modification

CHANGE SUMMARY.   The revised language of this provision mandates that all basements be provided with at least one emergency escape and rescue opening, except for those basements no more than 200 square feet in floor area used solely for housing mechanical equipment. Previously, the emergency openings were required only for basements containing habitable space.

2006 CODE: R310.1 Emergency Escape and Rescue Required. Basements ~~with habitable space~~ and every sleeping room shall have at least one operable emergency and rescue opening. Where basements contain one or more sleeping rooms, emergency egress and rescue openings shall be required in each sleeping room, but shall not be required in adjoining areas of the basement. Where emergency escape and rescue openings are provided they shall have a sill height of not more than 44 inches (1118 mm) above the floor. Where a door opening having a threshold below the adjacent ground elevation serves as an emergency escape and rescue opening and is provided with a bulkhead enclosure, the bulkhead enclosure shall comply with Section R310.3. The net clear opening dimensions required by this section shall be obtained by the normal operation of the emergency escape and rescue opening from the inside. Emergency escape and rescue openings with a finished sill height below the adjacent ground elevation shall be provided with a window well in accordance with Section R310.2. <u>Emergency escape and rescue openings shall open directly into a public way, or to a yard or court that opens to a public way.</u>

R310.1 continues

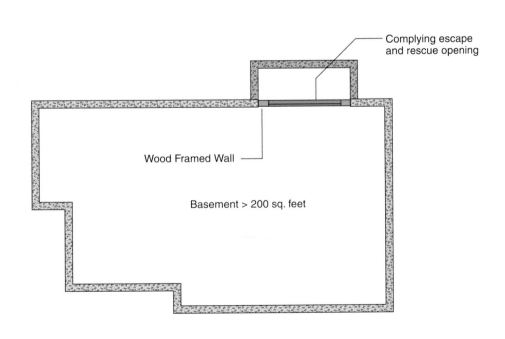

Complying escape and rescue opening

Wood Framed Wall

Basement > 200 sq. feet

Emergency Escape and Rescue Openings in Basements

R310.1 continued

Exception: <u>Basements used only to house mechanical equipment and not exceeding total floor area of 200 square feet (18.58 m²).</u>

CHANGE SIGNIFICANCE. Emergency escape and rescue openings are intended to provide a secondary means of escape under emergency conditions where the means of egress is unusable. In addition, such openings allow emergency responders and others an increased opportunity to rescue individuals who may be trapped within the dwelling. The requirement for emergency escape and rescue openings had previously been limited to sleeping rooms and basements containing habitable space. This code change maintains the provision for such openings in sleeping rooms, while extending the requirement to almost all basements. Only those basements that are no more than 200 square feet in floor area are exempt, provided they are used solely as mechanical equipment areas.

In many geographical areas a basement is a common feature of a house, often originally constructed and utilized in an unfinished condition. This frequently resulted in the basement being considered nonhabitable, permitting the omission of an emergency escape and rescue opening. Over time, basement remodeling would result in the partial or complete finish-out of the space. The modifications necessary to add a complying opening where none previously existed are expensive; thus, many times such openings were not provided. By mandating an emergency escape and rescue opening for virtually all basements at the time of initial construction, the code addresses the concern in a consistent and effective manner. The revised language also eliminates varying definitions of the term *habitable,* which will provide for uniform application of the provision.

Application of the exception is limited to those small basements intended for use only as mechanical equipment areas. Where a basement of 200 square feet or less in floor area is to be used as a storage area, for laundry facilities, or for a similar purpose, a complying emergency escape and rescue opening is mandated.

Language has also been added to the provision to clarify that escape or rescue must occur directly between the exterior and the opening. By mandating that travel through the opening be direct into a public way, or to a yard or court that opens to a public way, the code has been clarified to prohibit travel through intervening rooms or areas, and has eliminated such emergency egress to a court or area with no access to public way.

R310.1.4, R310.4

Operation of Emergency Escape and Rescue Openings

CHANGE TYPE. Addition

CHANGE SUMMARY. The limitations for operation of emergency escape and rescue openings, including any security devices that cover the openings, now include performance language preventing any operation that takes special knowledge.

2006 CODE: R310.1.4 Operational Constraints. Emergency escape and rescue openings shall be operational from the inside of the room without the use of keys, ~~or~~ tools, <u>or special knowledge.</u>

R310.4 Bars, Grills, Covers, and Screens. Bars, grills, covers, screens, or similar devices are permitted to be placed over emergency escape and rescue openings, bulkhead enclosures, or window wells that serve such openings, provided the minimum net clear opening size complies with Sections R310.1.1 to R310.1.3, and such devices shall be releasable or removable from the inside without the use of a key, tool, <u>special knowledge,</u> or force greater than that which is required for normal operation of the escape and rescue opening.

CHANGE SIGNIFICANCE. It is critical that a person be able to easily and quickly operate an emergency escape and rescue opening. The code has previously required such openings to be operational from the inside of the room without the use of keys or tools. The additional language now also requires that no special knowledge be necessary. Devices such as combination locks or latches that require a unique op-

R310.1.4, R310.4 continues

R310.1.4, R310.4 continued

eration sequence are now prohibited. These types of locking devices pose a distinct hazard in an emergency to anyone who is unfamiliar with their operation.

Bars, grills, and similar obstructions are occasionally placed over emergency escape and rescue openings for security purposes. Previously, the removal or release of such devices from the inside could not be precluded by the use of a key, tool, or unusual level of force. The devices must now also be openable from the inside without any special knowledge by the user, such as in the operation of the unlocking or unlatching hardware.

R310.5

Emergency Openings under Decks and Porches

CHANGE TYPE.   Addition

CHANGE SUMMARY.   A new section permits travel from an escape and rescue opening to a safe area to occur below an exterior deck or porch, provided the path is at least 36 inches in height.

2006 CODE: **310.5 Emergency Escape Windows under Decks and Porches.** Emergency escape windows are allowed to be installed under decks and porches provided the location of the deck allows the emergency escape window to be fully opened and provides a path not less than 36 inches (914 mm) in height to a yard or court.

CHANGE SIGNIFICANCE.   The provisions of Section R310.1 mandate that a window utilized for emergency escape and rescue be located directly adjacent to an open space such as a street, alley, yard, or court. However, the design of some houses, particularly townhouses, makes the underside of decks or porches the only location where such an opening can be located. This new provision permits emergency escape or rescue openings to lead directly to an under-deck or under-

R310.5 continues

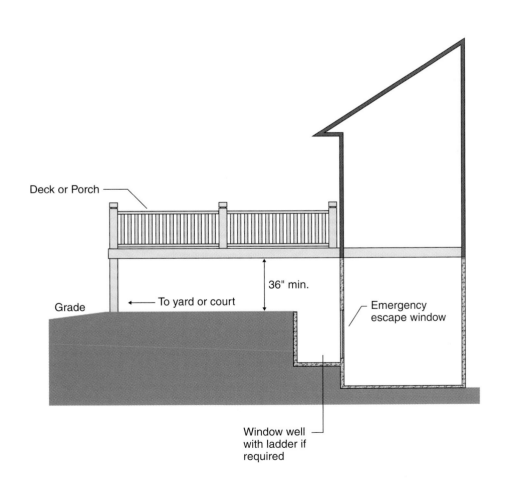

Deck or Porch

36" min.

Grade   ← To yard or court

Emergency
escape window

Window well
with ladder if
required

R310.5 continued   porch path having a minimum height of 36 inches. The 36-inch-height requirement is based on the minimum mandated size of a window well that serves an emergency escape and rescue opening.

It is expected that the emergency rescue or escape travel route below a porch or deck be consistent with the general requirements. The location of the porch or deck in relation to that of the window must continue to provide for the mandated net clear-opening area, height, and width. In addition, the path of travel under the porch or deck must lead directly to a yard or court.

CHANGE TYPE. Modification

CHANGE SUMMARY. A landing at an exterior door is now permitted to have a minimal slope for drainage purposes, limited to 1:48 maximum. In addition, the door may not swing over the stairway where a landing is not required.

2006 CODE: R311.4.3 Landings at Doors. There shall be a floor or landing on each side of each exterior door. <u>The floor or landing at the exterior door shall not be more than 1.5 inches (38 mm) lower than the top of the threshold. The landing shall be permitted to have a slope not to exceed 0.25 unit vertical in 12 units horizontal (2-percent).</u>

Exceptions:
1. Where a stairway of two or fewer risers is located on the exterior side of a door, other than the required exit door, a landing is not required for the exterior side of the door<u>, provided the door, other than an exterior storm or screen door, does not swing over the stairway.</u>

~~The floor or landing at the exit door required by Section R311.4.1 shall not be more than 1.5 inches (38 mm) lower than the top of the threshold. The floor or landing at exterior doors other than the exit door required by Section R311.4.1 shall not be required to comply~~

R311.4.3
Landings at Exterior Doors

R311.4.3 continues

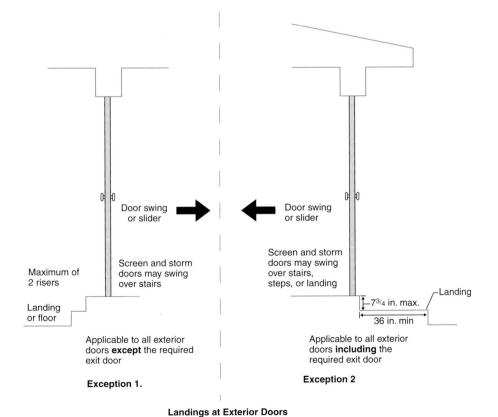

Door swing or slider → ← Door swing or slider

Maximum of 2 risers

Screen and storm doors may swing over stairs

Screen and storm doors may swing over stairs, steps, or landing

Landing or floor

7³/₄ in. max.

36 in. min

Landing

Applicable to all exterior doors **except** the required exit door

Applicable to all exterior doors **including** the required exit door

Exception 1.

Exception 2

Landings at Exterior Doors

R311.4.3 continued ~~with this requirement but shall have a rise no greater than that permitted in Section R311.5.3.~~

> ~~**Exception:**~~
>
> **2.** The <u>exterior</u> landing at an exterior doorway shall not be more than 7¾ inches (196 mm) below the top of the threshold, provided the door, other than an exterior storm or screen door, does not swing over the landing.
>
> **3.** <u>The height of floors at exterior doors other than the exit door required by section R311.4.1 shall not be more than 7¾ inches lower than the top of the threshold.</u>

The width of each landing shall not be less than the door served. Every landing shall have a minimum dimension of 36 inches (914 mm) measured in the direction of travel.

CHANGE SIGNIFICANCE. Previously, the code did not provide guidance as to the permitted slope of an exterior landing. It was generally assumed that the landing be level in order to provide a safe and stable walking surface. Additional language now allows a slope for drainage purposes but limits the slope to a maximum of 2%. This limitation is consistent with provisions in other codes that recognize some degree of slope is necessary while maintaining a relatively level landing surface.

The code has previously recognized that where a stairway consists of only one or two risers, a landing is not required at any doors other than the required exit door. The language has been modified to limit the application of this provision to only those situations where the door does not swing over the stairway. Since travel through the doorway will have an abrupt elevation change, the swing of the door over the stairway creates a more hazardous condition. Due to installation methods, the limitation does not apply to storm doors and screen doors.

The general format of the provisions relating to landings at exterior doors has also been modified to more clearly indicate the requirements. The previous language was found by many to be confusing, particularly in regard to what type of landing must be provided at the required exterior exit door. There are two options for addressing the required exit door: a landing no more than 1½ inches lower than the top of the threshold and a landing no more than 7¾ inches below the top of the threshold, where the door does not swing over the landing. A third option is available for doors other than the required exit door that allows for a stairway of one or two risers without a landing.

CHANGE TYPE. Clarification

CHANGE SUMMARY. Language has been added to clarify that a landing is not required at a flight of stairs between an attached garage and a dwelling where a door does not swing over the stairs.

2006 CODE: R311.5.4 Landings for Stairways. There shall be a floor or landing at the top and bottom of each stairway.

> **Exception:** A floor or landing is not required at the top of an interior flight of stairs, <u>including stairs in an enclosed garage,</u> provided a door does not swing over the stairs.

A flight of stairs shall not have a vertical rise greater than 12 feet (3658 mm) between floor levels or landings.

The width of each landing shall not be less than the stairway served. Every landing shall have a minimum dimension of 36 inches (914 mm) measured in the direction of travel.

CHANGE SIGNIFICANCE. Stairs between an attached garage and a dwelling are now specifically identified as interior stairs for the purpose of landing requirements. Consistent with allowances for other stairs within the dwelling, a landing is not required at the top of the

R311.5.4 continues

R311.5.4

Landings at Garage Stairways

R311.5.4 continued   stair flight, provided a door does not swing out over the stairs. In some cases, the garage stairs were previously considered exterior stairs, as they were viewed as being outside of the dwelling portion of the structure, and such stairs were often required to be provided with a complying landing on the garage side of the door to the dwelling. The new language recognizes that the stairs between the garage and the dwelling should be considered interior stairs.

R311.6.1
Maximum Slope of Ramps

CHANGE TYPE.   Modification

CHANGE SUMMARY.   The revised language in the 2006 code requires that ramps, when installed, must have a maximum slope of 1 unit vertical to 12 units horizontal. This provides for a more level ramp than the 2003 code slope requirement of 1 unit vertical to 8 units horizontal.

2006 CODE: R311.6.1 Maximum Slope. ~~Ramps shall have a maximum slope of one unit vertical in eight units horizontal (12.5-percent slope).~~ Ramps shall have a slope of one unit vertical in 12 units horizontal (8.3-percent slope).

> **Exception:** Where it is technically infeasible to comply because of site constraints, ramps may have a maximum slope of one unit vertical to eight horizontal (12.5 percent slope).

CHANGE SIGNIFICANCE.   The maximum slope of ramps for buildings regulated under the IRC has been modified to be more consistent with the accessibility ramps in the IBC by deleting the 2003 code maximum slope of 1 unit vertical to 8 units horizontal (12.5% slope) and inserting a maximum slope of 1 unit vertical to 12 units horizontal (8.3% slope). The main justification provided for this change is the issue of accessibility for the aging population and the need to install ramps that are us-

R311.6.1 continues

R311.6.1 continued   able by the elderly. Even though most buildings constructed under the IRC are not required to be accessible and hence the installation of ramps is not mandatory, when installed, all ramps must now comply with the new and lower slope of 8.3%. As access ramps or sloped walkways are the modification most often needed by persons who use wheelchairs or people who use canes or walkers, and for such users a ramp with a slope of 1:8 is not usable, the new maximum 1:12 slope will address the need of a growing segment of our population.

CHANGE TYPE. Clarification

CHANGE SUMMARY. The code section regulating guards has been expanded to clarify that elevated ramps are subject to the provisions.

2006 CODE: R312.1 Guards. Porches, balconies, ramps or raised floor surfaces located more than 30 inches (762 mm) above the floor or grade below shall have guards not less than 36 inches (914 mm) in height. Open sides of stairs with a total rise of more than 30 inches (762 mm) above the floor or grade below shall have guards not less than 34 inches (864 mm) in height measured vertically from the nosing of the treads.

Porches and decks which are enclosed with insect screening shall be provided with guards where the walking surface is located more than 30 inches (762 mm) above the floor or grade below.

CHANGE SIGNIFICANCE. Porches, balconies, and stairs have previously been identified as requiring guards where they are located more than 30 inches above the surface below. Although ramps were generally considered as walking surfaces also regulated by the provisions, the addition of ramps to the code language clarifies the intent. The code now specifically identifies that where a ramp is elevated above the grade or floor below more than 30 inches, a minimum 30-inch-high guard is required.

R312.1

Guards at Elevated Ramps

R313.1.1

Smoke Alarms in Existing Dwellings

CHANGE TYPE.   Modification

CHANGE SUMMARY.   Code language has been added to reflect that smoke alarms are not required in existing dwellings where the work is limited to window or door replacement or when an exterior porch or deck is constructed.

2006 CODE: R313.1.1 Alterations, Repairs and Additions. When ~~interior~~ alterations, repairs or additions requiring a permit occur, or when one or more sleeping rooms are added or created in existing dwellings, the individual dwelling unit shall be provided with smoke alarms located as required for new dwellings; the smoke alarms shall be interconnected and hard wired.

Exceptions:

1. (no change to text)
2. ~~Repairs to~~ <u>Work involving</u> the exterior surfaces of dwellings, such as the replacement of roofing or siding, <u>or the addition or replacement of windows or doors, or the addition of a porch or deck,</u> are exempt from the requirements of this section.

CHANGE SIGNIFICANCE.   As a general requirement, dwelling units undergoing alterations, additions, or repairs must be provided with smoke alarms in those locations mandated for new construction. An exception has previously exempted dwellings from this requirement where the work was limited to exterior surfaces, such as where roofing or siding replacement was done. Additional language now indicates that smoke alarms are also not required where the work includes only construction of an exterior deck or porch, or where exterior doors or windows are being replaced or added.

CHANGE TYPE.   Addition

CHANGE SUMMARY.   An additional method of separating dwelling units in a two-family dwelling now allows for fire-resistive protection at the ceiling line of each unit rather than at the wall line in the attic.

2006 CODE: **R317.1 Two-Family Dwellings.**   Dwelling units in two-family dwellings shall be separated from each other by wall and/or floor assemblies having not less than a 1-hour fire-resistance rating when tested in accordance with ASTM E 119. Fire-resistance-rated floor-ceiling and wall assemblies shall extend to and be tight against the exterior wall, and wall assemblies shall extend to the underside of the roof sheathing.

Exceptions:

1. A fire-resistance rating of ½ hour shall be permitted in buildings equipped throughout with an automatic sprinkler system installed in accordance with NFPA 13.

2. Wall assemblies need not extend through attic spaces when the ceiling is protected by not less than ⅝-inch (15.9 mm) Type X gypsum board and an attic draftstop constructed as specified in Section R502.12.1 is provided above and along the wall assembly separating the dwellings. The structural framing supporting the ceiling shall also be protected by not less than ½-inch (12.7 mm) gypsum board or equivalent.

R317.1 continues

R317.1
Fire Separation of Two-Family Dwellings

Attic draftstop constructed per Section R502.12.1

Dwelling Unit

Dwelling Unit

Ceiling membrane to be minimum ⅝-inch type X gypsum board

Minimum 1-hour FRR wall

R317.1 continued

CHANGE SIGNIFICANCE.   In the past, the code mandated that the attic space above each dwelling unit in a two-family dwelling be separated by a minimum 1-hour fire-resistance-rated wall assembly. This approach created a complete vertical fire-separation element isolating one unit from the other. A new exception provides for an alternative method of separating units, using a combination of both ceiling protection and attic draftstopping.

Certain methods of roof framing sometimes make it difficult to extend the 1-hour wall through the attic space. Framing perpendicular to the unit-separation-wall line, the installation of purlins, and attic framing bracing are common conditions that make it difficult to construct a complete 1-hour attic separation. For these and other reasons, a second method is now available that provides for a minimum of $5/8$-inch Type X gypsum board on the ceiling and construction of the attic separation as a draftstop.

Where the separation is accomplished by installing the required gypsum board on the ceiling, the structural framing supporting the ceiling must also be protected. However, the code is silent on the means of dealing with penetrations and other openings that may occur in the ceiling membrane. There is an expectation that any interruption of the gypsum board be addressed by a method that maintains the level of fire-separation set forth in the code. The intent of the new alternative is to provide an equivalent level of separation to that provided by the 1-hour wall separation.

CHANGE TYPE. Modification

CHANGE SUMMARY. The provisions addressing decay damage have been modified to apply uniformly to all geographic areas. A definition for naturally durable wood has been included, and American Wood Preservers' Association (AWPA) Standard U1 replaces the former C (commodity) standards that were previously referenced.

2006 CODE: **R319.1 Location Required.** ~~In areas subject to decay damage as established by Table R301.2(1),~~ Protection from decay shall be provided in the following locations ~~shall require~~ by the use of ~~an approved species and grade of lumber,~~ naturally durable wood or wood that is pressure preservatively treated in accordance with AWPA ~~C1, C2, C3, C4, C9, C15, C18, C22, C23, C24, C28, C31, C33, P1, P2 and P3, or decay-resistant heartwood of redwood, black locust, or cedars.~~ U1 for the species, product, preservative, and end use. Preservatives shall be listed in Section 4 of AWPA U1.

1. through 7. (no change to text)

R319.1, R202 continues

Moderate to Severe

Slight to Moderate

None to Slight

**Decay Probability Map Has Now Been
Deleted From The Code**

R319.1, R202 continued

R319.1.1 Field Treatment. <u>Field cut ends, notches, and drilled holes of preservatively treated wood shall be treated in the field in accordance with AWPA M4.</u>
(subsequent sections renumbered)

R202 Definitions.

■ <u>**NATURALLY DURABLE WOOD.** The heartwood of the following species: decay-resistant redwood, cedars, black locust and black walnut.</u>

<u>Note: Corner sapwood is permitted if 90 percent or more of the width of each side on which it occurs is heartwood.</u>
~~Figure R301.2(7) Decay Probability Map~~
(Subsequent figures renumbered)

CHANGE SIGNIFICANCE. Previously, the code text indicated that the degree of decay damage differed from one geographical area to another, creating questions as to the applicability of the provisions addressing protection against decay. While the general climate of one area may be more conducive to outdoor decay conditions in general than that in another climatic area, in reality decay conditions within structures are more likely to exist because of drainage problems, leaks, workmanship deficiencies, landscape watering, lack of ventilation, and similar conditions rather than because of climate. Therefore, when the conditions listed in Section 319 are present, decay protection is needed regardless of geographic location. As a result, the modified code language now addresses protection against decay equally for all geographic areas, eliminating the need for the Decay Probability Map [2003 IRC Figure R301.2(7)].

Provisions have been added for the field treatment of preservatively treated wood that are consistent with those of Section R320.3.1 for protection against termites. In addition, a new definition of naturally durable wood is provided that includes those woods previously mentioned as decay-resistant.

R319.1.5

Protection of Glued-Laminated Members against Decay

CHANGE TYPE.   Addition

CHANGE SUMMARY.   Requirements have been added to the code for the use of structural glued laminated timbers in exterior locations.

2006 CODE: <u>**319.1.5 Exposed Glued-Laminated Timbers.**</u>   <u>The portions of glued-laminated timbers that form the structural supports of a building or other structure and are exposed to weather and not properly protected by a roof, eave, or similar covering shall be pressure treated with preservative, or be manufactured from naturally durable or preservative-treated wood.</u>

CHANGE SIGNIFICANCE.   In order to provide for adequate protection against decay, the code has long addressed the use of wood members in exterior locations. New provisions now also specifically regulate glued laminated wood timbers that are exposed to the weather without proper roof protection. The required methods of protection are consistent with those for other wood members regulated for protection against decay, including the use of preservative-treated wood in the manufacture of the timbers.

R320, Table R301.2(1)

Protection against Subterranean Termites

CHANGE TYPE. Clarification

CHANGE SUMMARY. The provisions for addressing damage caused from termites have been rewritten to reflect current practices and terminology. The change also simplifies the requirements for termite protection and allows individual jurisdictions to determine whether such protective measures are needed.

2006 CODE: **Section R320 Protection Against Subterranean Termites R320.1 Subterranean Termite Control Methods.** In areas ~~favorable~~ subject to ~~termite~~ damage from termites as ~~established~~ indicated by Table R301.2(1), methods of protection shall be one of the following methods or a combination of these methods:

1. ~~by c~~Chemical ~~soil~~ termiticide treatment, as provided in Section R320.2~~.~~.

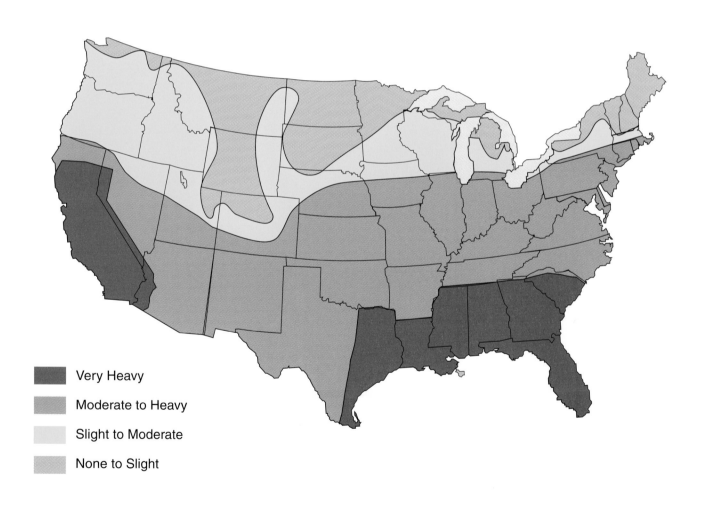

Very Heavy

Moderate to Heavy

Slight to Moderate

None to Slight

Termite Infestation Probablity Map

2. Termite baiting system installed and maintained according to the label.

3. ~~p~~Pressure preservatively treated wood in accordance with the AWPA standards listed in Section R319.1~~.~~.

4. ~~n~~Naturally termite-resistant wood as provided in Section R320.3.

5. ~~or p~~Physical barriers as provided in Section R320.4 ~~(such as metal or plastic termite shields), or any combination of these methods.~~

R320.1.1 Quality Mark.   (no changes to text)

R320.1.2 Field Treatment.   Field cut ends, notches, and drilled holes of pressure preservatively treated wood shall be retreated in the field in accordance with AWPA M4.

R320.2 Chemical ~~Soil~~ Termiticide Treatment.   Chemical termiticide treatment shall include soil treatment and/or field applied wood treatment. The concentration, rate of application, and ~~treatment~~ method of treatment of the chemical termiticide shall be ~~consistent~~ in strict accordance with ~~and never less than~~ the termiticide label.

R320.3 ~~Pressure Preservatively Treated and n~~Naturally Resistant Wood.   Heartwood of redwood and eastern red cedar shall be considered termite resistant. ~~Pressure preservatively treated wood and naturally termite-resistant wood shall not be used as a physical barrier unless a barrier can be inspected for any termite shelter tubes around the inside and outside edges and joints of a barrier.~~

R320.4 Barriers.   Approved physical barriers, such as metal or plastic sheeting or collars specifically designed for termite prevention, shall be installed in a manner to prevent termites from entering the structure. Shields placed on top of an exterior foundation wall are permitted to be used only if in combination with another method of protection.

~~R320.4~~ R320.5 Foam Plastic Protection.   (no changes to text)

Table R301.2(1) Climatic and Geographic Design Criteria
(no change to contents of table)

a. and b. (no change to text)

c. The jurisdiction shall fill in this part of the table ~~with "very heavy," "moderate to heavy," "slight to moderate," or "none to slight" in accordance with Figure R301.2(6)~~ to indicate the need for protection depending on whether there has been a history of local subterranean termite damage.

d. through k. (no change to text)

R320, Table R301.2(1) continues

R320, Table R301.2(1) continued

CHANGE SIGNIFICANCE. Although the format of Section R320 has been substantially revised, there are no significant changes to the technical content. The title was revised to reflect that the provisions are intended to address the control of subterranean termites. Allowing for increased identification, the five acceptable methods of protection are now individually listed. In addition, the provisions for physical barriers have been expanded by defining the term and clarifying the use of termite shields.

The previous use of terms on Figure R301.2(6) implied that there were varying degrees of termite protection, depending on the level of termite-infestation probability. The revised language clarifies that historical data are to be used by the jurisdiction in the determination of whether termite protection is required, with no need to reference Figure R301.2(6). The probability classifications of the table remain relevant to application of the provisions of Section R320.5 for foam-plastic protection.

CHANGE TYPE.   Clarification

CHANGE SUMMARY.   In the establishment of the design flood elevation for a building constructed in a flood hazard area, new language clarifies that the building official can require the applicant to use data available from other sources where such elevations are not specified.

2006 CODE: R324.1.3.1 Determination of Design Flood Elevations.   If design flood elevations are not specified, the building official is authorized to require the applicant to:

1. Obtain and reasonably utilize data available from a federal, state, or other source, or

2. Determine the design flood elevation in accordance with accepted hydrologic and hydraulic engineering practices used to define special flood hazard areas. Determinations shall be undertaken by a registered design professional who shall document that the technical methods used reflect currently accepted engineering practice. Studies, analyses, and computations shall be submitted in sufficient detail to allow thorough review and approval.

R324.1.3.1 continues

R324.1.3.1
Determination of Design Flood Elevations

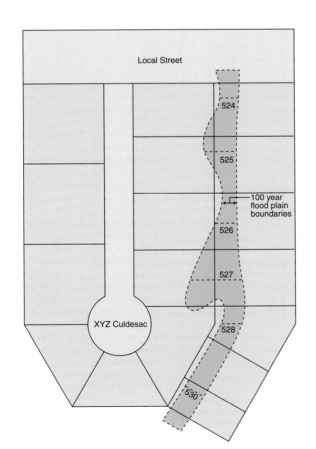

R324.1.3.1 continued

CHANGE SIGNIFICANCE. Buildings constructed in flood hazard areas shall be assigned a design flood elevation for the proper application of flood-resistant construction requirements. In some cases, the design flood elevation is not specified for flood hazard areas that are delineated on flood hazard maps prepared by the National Flood Insurance Program. This code change authorizes the building official to require the applicant to use data available from other sources, which may be based on existing studies or surveys. Alternatively, the determination may be achieved through accepted hydrologic and hydraulic engineering techniques. The provision does not preclude the building official from providing design-flood-elevation information to an applicant when such information is available. It is implicit that the building official review the submission of all necessary data to ensure it is adequate.

CHANGE TYPE. Modification

CHANGE SUMMARY. The surface drainage requirements of the lot have been clarified, and specific slope criteria for swales have been added.

2006 CODE: R401.3 Drainage. Surface drainage shall be diverted to a storm sewer conveyance or other approved point of collection so as to not create a hazard. Lots shall be graded so as to drain surface water away from foundation walls. The grade shall fall a minimum of 6 inches (152 mm) within the first 10 feet (3048 mm).

> **Exception:** Where lot lines, walls, slopes or other physical barriers prohibit 6 inches (152 mm) of fall within 10 feet (3048 mm), the final grade shall slope away from the foundation at a minimum slope of 5% and the water shall be directed to drains or swales ~~shall be provided~~ to ensure drainage away from the structure. Swales shall be sloped a minimum of 2% when located within 10 feet (3048 mm) of the building foundation. Impervious surfaces within 10 ft (3048 mm) of the building foundation shall be sloped a minimum of 2% away from the building.

CHANGE SIGNIFICANCE. The 2003 IRC exception requires only that drains or swales be provided in the event that 6 inches of fall within 10 feet (5% slope) is not attainable and prescriptive slope requirements for this condition are unclear, but no further criteria for drainage slope are provided. This lack of specific criteria has caused

R401.3 continues

R401.3
Drainage

R401.3 continued   inconsistency and some confusion in the application of drainage requirements. This modification and the additional criteria for swale slopes clarify the intent of the code by requiring a 5% minimum slope away from the structure, especially when 10 feet is not possible. It also addresses the use of swales to convey surface water between closely spaced buildings in "zero lot line," "cluster," or other development practices that are increasingly encouraged by many jurisdictions.

CHANGE TYPE.   Modification

R401.4.2
Compressive or Shifting Soil

CHANGE SUMMARY.   The modification relocates Section R401.5 to Section 401.4.2 and revises the technical content to allow the geotechnical analysis to determine whether compressible or shifting soils need to be completely removed before the placement of foundation.

2006 CODE: ~~R401.5~~ <u>R401.4.2</u> Compressible or Shifting Soil. <u>Instead of a complete geotechnical evaluation,</u> when top or subsoils are compressible or shifting, such soils shall be removed to a depth and width sufficient to assure stable moisture content in each active zone and shall not be used as fill or stabilized within each active zone by chemical, dewatering, or presaturation.

CHANGE SIGNIFICANCE.   Within the current format of the 2003 IRC, the requirements for compressible or shifting soils are found in Section 401.5, which is a distinct and separate section from that of soil tests, found in Section R401.4. By application of the 2003 IRC Section R401.4, the building official must determine whether to require a soils test "in areas likely to have expansive, compressible, shifting or other unknown soil characteristics." These two separate and distinct requirements created a situation where the geotechnical investigation and the soils report could provide analysis and recommendations for compressible and shifting soils that possibly could not be used, be-

R401.4.2 continues

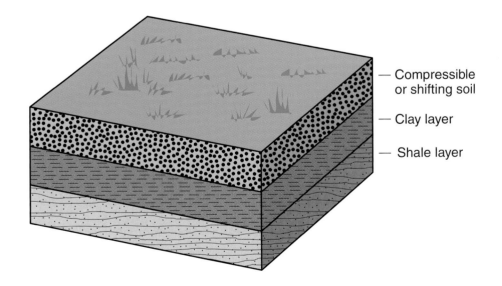

— Compressible or shifting soil

— Clay layer

— Shale layer

Either complete geotechnical report to determine what to do with the compressible or shifting soil or remove it.

Compressible or shifting soils are those that are subject to volume change or being easily compressed which can cause foundation settlement.

R401.4.2 continued   cause Section R401.5 required the removal of such soils regardless of the geotechnical report recommendations. The 2006 IRC has resolved this problem by relocating the criteria for compressible and shifting soils as a subsection of the soil tests section. The revision further clarifies that the removal of compressible or shifting soils is mandated only when there is no geotechnical evaluation. When there is a geotechnical evaluation, the resulting recommendations are to be used. Further reasons for this revision are that many soil-investigation reports classify bearing soils as compressible and that design procedures exist for designing foundations on compressible soils (e.g., Post-Tensioning Institute [PTI] Design and Construction of Post-Tensioned Slabs-On-Ground). Thus, it is not necessary for all compressible soils to be removed; rather, the geotechnical engineer must decide the best course of action to deal with any compressible soils.

Table R402.2

Minimum Specified Compressive Strength of Concrete

CHANGE TYPE. Addition

CHANGE SUMMARY. A new footnote, "f," allows the reduction of air entrainment to 3% for garage floors when the minimum concrete compressive strength is at least 4000 psi.

2006 CODE:

TABLE R402.2   Minimum Specified Compressive Strength of Concrete

Type or Locations of Concrete Construction	Minimum Specified Compressive Strength[a] (f'_c) Weathering Potential[b]		
	Negligible	Moderate	Severe
Basement walls, foundations, and other concrete not exposed to the weather	2500	2500	2500[c]
Basement slabs and interior slabs on grade, except garage floor slabs	2500	2500	2500[c]
Basement walls, foundation walls, exterior walls, and other vertical concrete work exposed to the weather	2500	3000[d]	3000[d]
Porches, carport slabs and steps exposed to the weather, and garage floor slabs	2500	3000[d,e,f]	3500[d,e,f]

For SI: 1 pound per square inch = 6.895 KPa.

a. At 28 days psi.
b. See Table R301.2(1) for weathering potential.
c. Concrete in these locations that may be subject to freezing and thawing during construction shall be air-entrained concrete in accordance with Footnote d.
d. Concrete shall be air-entrained. Total air content (percent by volume of concrete) shall not be less than 5% or more than 7%.
e. See Section R402.2 for ~~minimum cement~~ maximum cementitious materials content.
f. For garage floors with a steel troweled finish, the total air content (percent by volume of concrete) is permitted to be reduced to not less than 3% if the specified compressive strength of the concrete is increased to not less than 4000 psi.

CHANGE SIGNIFICANCE. The revision allows air entrainment in moderate and severe weathering regions to be reduced to 3% when concrete compressive strength is increased from 3000 or 3500 psi to at least 4000 psi. The purpose of the air-entrainment is to protect the slab from freeze-thaw deterioration aggravated by de-icer chemicals that may be carried into the garage on vehicle tires and the underside of the vehicle.

Most homeowners prefer a garage slab with a smooth trowel finish to facilitate cleaning. Experience has shown that it is very difficult to get a smooth, durable finish by using a steel trowel on concrete with total air content in the range of 5% to 7%. This change provides an option to address the problem in severe and moderate weathering regions. Field ex-

Table R402.2 continues

Table R402.2 continued   perience has shown that durable concrete may be obtained with a lesser amount of air-entrainment if the specified compressive strength of the concrete is increased to 4000 psi. The higher concrete strength is accompanied by a denser cement-paste matrix, which results in the concrete being less permeable.

CHANGE TYPE.   Modification

CHANGE SUMMARY.   The required depth of interior footings is reduced from 18 inches to 12 inches below the top of the slab.

2006 CODE: R403.1.4.2 Seismic Conditions.   In Seismic Design Categories $\underline{D_0}$, D_1, and D_2, interior footings supporting bearing or bracing walls and cast monolithically with a slab on grade shall extend to a depth of not less than ~~18 inches (457 mm)~~ 12 inches (305mm) below the top of the slab.

CHANGE SIGNIFICANCE.   Section R403.1.4 requires a depth of 12 inches below the undisturbed ground for exterior footings in all SDCs. Comparing this requirement with the 2003 code Section R403.1.4.2 requirement for the depth of interior footings in Seismic Design Categories D_1 and D_2 reveals that interior footings are required to be placed deeper than exterior footings in cases where undisturbed soil is at grade or within 6 inches of the grade. The 2006 code's reduction in the depth of interior footings cast monolithically with a slab on grade provides that interior footings will still be placed at a depth where adequate bearing capacity is provided but the interior footings are not required to extend to a point that is deeper than the exterior footings.

R403.1.4.2
Seismic Conditions

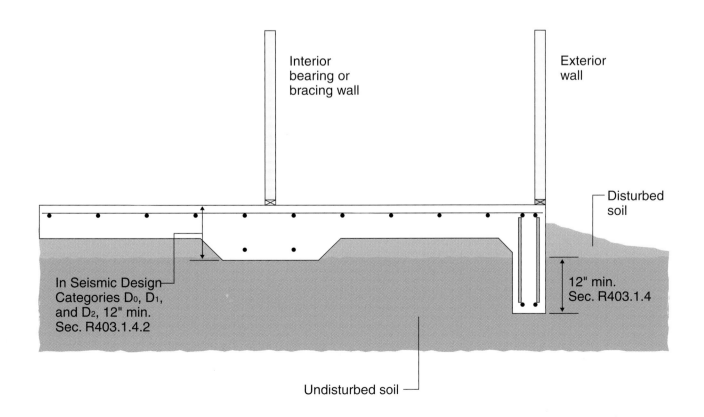

Interior bearing or bracing wall

Exterior wall

Disturbed soil

In Seismic Design Categories D₀, D₁, and D₂, 12" min. Sec. R403.1.4.2

12" min. Sec. R403.1.4

Undisturbed soil

R403.1.6.1

Foundation Anchorage in Seismic Design Categories C, D_0, D_1, and D_2

CHANGE TYPE. Modification

CHANGE SUMMARY. The washer size requirement of 3 inches by 3 inches by ¼ inch has been revised to "properly sized cut washers" for those walls that do not contain braced wall panels.

2006 CODE: R403.1.6.1 Foundation Anchorage in Seismic Design Categories C, $\underline{D_0}$, D_1, and D_2. In addition to the requirements of Section R403.1.6, the following requirements shall apply to wood light-frame structures in Seismic Design Categories $\underline{D_0}$, D_1, and D_2 and wood light-frame townhouses in Seismic Design Category C.

1. Plate washers conforming to Section R602.11.1 shall be ~~used on each bolt~~ <u>provided for all anchor bolts over the full length of required braced wall lines. Properly sized cut washers shall be permitted for anchor bolts in wall lines not containing braced wall panels.</u>

2. Interior braced wall plates shall have anchor bolts spaced at not more than 6 feet (1829mm) on center and located within 12 inches (305 mm) from the ends of each plate section when supported on a continuous foundation.

3. Interior bearing wall sole plates shall have anchor bolts spaced at not more than 6 feet (1829 mm) on center and located within 12 inches (305mm) from the ends of each plate section when supported on a continuous foundation.

4. The maximum anchor bolt spacing shall be 4 feet (1219mm) for buildings over two stories in height.

5. Stepped cripple walls shall conform to Section R602.11.3.

6. Where continuous wood foundations in accordance with Section R404.2 are used, the force transfer shall have a capacity equal to or greater than the connections required by Section R602.11.1 or the braced wall panel shall be connected to the

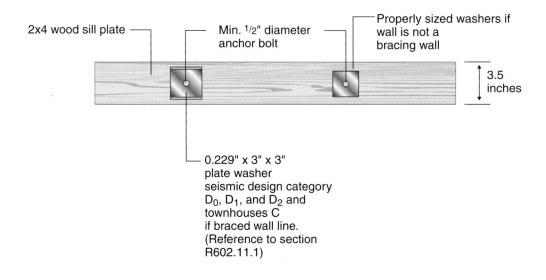

2x4 wood sill plate

Min. ½" diameter anchor bolt

Properly sized washers if wall is not a bracing wall

3.5 inches

0.229" x 3" x 3" plate washer seismic design category D_0, D_1, and D_2 and townhouses C if braced wall line. (Reference to section R602.11.1)

wood foundations in accordance with the braced wall panel-to-floor fastening requirements of Table 602.3(1).

CHANGE SIGNIFICANCE.   The new language allows "properly sized washers" in conventional construction practices rather than the 2003 requirement of 3 inches by 3 inches by ¼ inch found in Section R602.11.1, which is referenced from Section R403.1.6.1. This will allow for more flexible construction tolerances and eliminates the overly restrictive plate-washer requirements. The three inch-by-three inch washers create a construction and inspection problem: with a two-by-four plate, the bolt must be almost perfectly centered in the plate to avoid interference with the sheathing or the gypsum board.

A proposed code change in the 2000 to 2003 code development cycles increased the size of the plate washers from 2 inches by 2 inches by $\frac{3}{16}$ inch to 3 inches by 3 inches by ¼ inch in conventional construction. The 2000 NEHRP *Recommended Provisions,* Sections 12.4.2 and 12.5.3, impose the requirement for large plate washers only on shear walls and on braced wall lines. In conventional construction, the forces are substantially lower than those in which the tests produced the sill plate failures. The 2½ inches by 2½ inches by $\frac{3}{16}$ inch washers will adequately protect the sill plates against failure in the smaller, conventional loads.

R404.1

Concrete and Masonry Foundation Walls

CHANGE TYPE. Addition

CHANGE SUMMARY. New prescriptive criteria have been introduced for laterally supported foundation walls so that such foundation walls would not require design by licensed professionals unless required by the state law where the project is located.

2006 CODE: R404.1 Concrete and Masonry Foundation Walls. Concrete and masonry foundation walls shall be selected and constructed in accordance with the provisions of this section or in accordance with ACI 318, NCMA TR68–A or ACI 530/ASCE 5/TMS 402 or

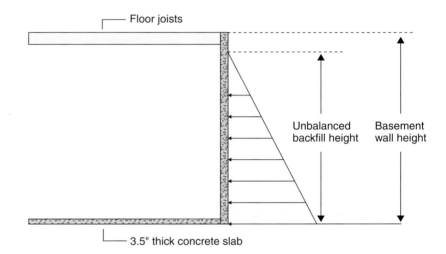

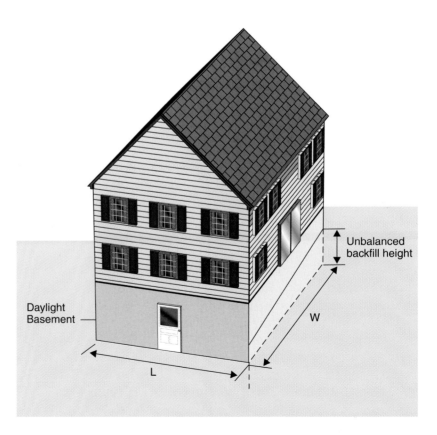

other approved structural standards. When ACI 318 or ACI 530/ASCE 5/TMS 402 or the provisions of this section are used to design concrete or masonry foundation walls, project drawings, typical details and specifications are not required to bear the seal of the architect or engineer responsible for design, unless otherwise required by the state law of the jurisdiction having authority.

Foundation walls that meet all of the following shall be considered laterally supported:

1. Full basement floor shall be minimum 3.5 inches (89 mm) thick concrete slab poured tight against the bottom of the foundation wall.

2. Floor joists and blocking shall be connected to sill plate at top of wall by the prescriptive method called out in Table R404.1(1), or shall be connected with an approved connector with listed capacity meeting Table R404.1(1).

3. Bolt spacing for the sill plate shall be spaced no greater than per Table R404.1(2).

4. Floor shall be blocked perpendicular to the floor joists. Blocking shall be full depth within 2 joist spaces of the foundation wall, and be flat-blocked with minimum 2-inch by 4-inch (51 mm by 102 mm) blocking elsewhere.

5. Where foundation walls support unbalanced load on opposite sides of the building, such as a daylight basement, the building aspect ratio, L/W, shall not exceed the value specified in Table R404.1(3). For such foundation walls, the rim board shall be attached to the sill with a 20 gage metal angle clip at 24 inches (610 mm) on center, with 5–8d nails per leg, or an approved connector supplying 230 plf (3.36 kN/m) capacity.

CHANGE SIGNIFICANCE. New prescriptive provisions have been added to Section R404.1 in the 2006 code, with limitation of supporting lateral soil loads only. The 2003 code provisions do not directly take into account the effects of gravity loads on the foundation wall. Inevitably, a situation might arise where an axial load is imposed on a foundation wall constructed in accordance with one of the prescriptive designs contained in IRC Tables R404.1.1(1) through R404.1.1(4) that would create a structurally unsafe condition. In fact, there are documented cases of failures of inadequately supported basement walls. The new approach in the 2006 code for providing lateral support follows a similar concept found in SSTD-10 and SSTD-13, which contain similar lateral support details and use the same wall thickness and reinforcement tables as the IRC. These standards show required floor blocking as described in this new code language and show metal clips bolted to sill and joists for lateral support. The lateral forces in the new Table R404.1(1) are provided in the code, with the expectation that given the required force, third-party vendors might develop connectors that can be used in place of the prescriptive metal clips.

The capacity of the 5–8d-nail connection has been taken from the table in the Wood Framed Construction Manual Commentary, ad-

R404.1 continues

R404.1 continued   justed for nonwind conditions. Capacities for the ¼″ steel angle clips were taken from the NF&PA's National Design Specifications (NDS).

Daylight basements create another problem. Rather than the forces directly transferring across the floor through the blocking, the unbalanced portion of the floor must be transferred to the perpendicular walls by shear. The aspect ratio limitation is due to the shear capacity of plywood nailed at 6 inches on center. Normal connection of the rim

TABLE R404.1(1)   Top Reactions and Prescriptive Support for Foundation Walls[a]

| | | Horizontal Reaction to Top (plf) | | |
| | | Soil Classes (Letter indicates connection types[b]) | | |
Maximum Wall Height (feet)	Maximum Unbalanced Backfill Height (feet)	GW, GP, SW, and SP Soils	GM, GC, SM-SC, and ML Soils	SC, MH, ML-CL, and Inorganic CL Soils
7	4	45.7 A	68.6 A	91.4 A
	5	89.3 A	133.9 B	178.6 B
	6	154.3 B	231.4 C	308.6 C
	7	245.0 C	367.5 C	490.0 D
8	4	40.0 A	60.0 A	80.0 A
	5	78.1 A	117.2 B	156.3 B
	6	135.0 B	202.5 B	270.0 C
	7	214.0 B	321.6 C	428.8 C
	8	320.0 C	480.0 C	640.0 D
9	4	35.6 A	53.3 A	71.1 A
	5	69.4 A	104.2 B	138.9 B
	6	120.0 B	180.0 B	240.0 C
	7	190.6 B	285.8 C	381.1 C
	8	284.4 C	426.7 C	568.9 D
	9	405.0 C	607.5 D	810.0 D

For SI: 1 foot = 304.8 mm, 1 pound = 0.454 kg, 1 plf = pounds per linear foot = 1.488 kg/m.

a. Loads are pounds per linear foot of wall. Prescriptive options are limited to maximum joist and blocking spacing of 24 inches on center.
b. Prescriptive Support Requirements:

Type	Joist/blocking Attachment Requirement
A	3–8d per joist per Table R602.3(1).
B	1–20 gage angle clip each joist with 5–8d per leg.
C	1¼-inch thick steel angle. Horizontal leg attached to sill bolt adjacent to joist/blocking, vertical leg attached to joist/blocking with ½-inch minimum diameter bolt.
D	2¼-inch thick steel angles, one on each side of joist/blocking. Attach each angle to adjacent sill bolt through horizontal leg. Bolt to joist/blocking with ½-inch minimum diameter bolt common to both angles.

TABLE R404.1.(2)   Maximum Plate Anchor-Bolt Spacing for Supported Foundation Wall[a]

| Maximum Wall Height (feet) | Maximum Unbalanced Backfill Height (feet) | Anchor Bolt Spacing (inches) | | |
| | | Soil Classes | | |
		GW, GP, SW, and SP Soils	GM, GC, SM-SC, and ML Soils	SC, MH, ML-CL, and Inorganic CL Soils
7	4	72	58	43
	5	44	30	22
	6	26	17	13
	7	16	11	8
8	4	72	66	50
	5	51	34	25
	6	29	20	15
	7	18	12	9
	8	12	8	6
9	4	72	72	56
	5	57	38	29
	6	33	22	17
	7	21	14	10
	8	14	9	7
	9	10	7	5

For SI: 1 inch = 25.4 mm, 1 foot = 304.8 mm.

a. Spacing is based on ½-inch diameter anchor bolts. For ⅝-inch diameter anchor bolts, spacing may be multiplied by 1.27, with a maximum spacing of 72 inches.

board to plate is not as strong as the allowable plywood shear. In order to get the capacity to a similar magnitude, additional clips are required. The new 2006 code provisions will provide for prescriptive design for most daylight basements for most residential buildings constructed under the IRC.

TABLE R404.1.(3)   Maximum Aspect Ratio, L/W for Unbalanced Foundations

| MaximumWall Height (feet) | Maximum Unbalanced Backfill Height (feet) | Soil Classes | | |
		GW, GP, SW, and SP Soils	GM, GC, SM-SC, and ML Soils	SC, MH, ML-CL, and Inorganic CL Soils
7	4	4.0	4.0	4.0
	5	4.0	3.4	2.6
	6	3.0	2.0	1.5
	7	1.9	1.2	0.9
8	4	4.0	4.0	4.0
	5	4.0	3.9	2.9
	6	3.4	2.3	1.7
	7	2.1	1.4	1.1
	8	1.4	1.0	0.7
9	4	4.0	4.0	4.0
	5	4.0	4.0	3.3
	6	3.8	2.6	1.9
	7	2.4	1.6	1.2
	8	1.6	1.1	0.8
	9	1.1	0.8	0.6

For SI: 1 foot = 304.8 mm.

R404.5

Retaining Walls

CHANGE TYPE.   Addition

CHANGE SUMMARY.   New section provides requirements for retaining walls.

2006 CODE: R404.5 Retaining Walls.   Retaining walls that are not laterally supported at the top and that retain in excess of 24 inches (610mm) of unbalanced fill shall be designed to ensure stability against overturning, sliding, excessive foundation pressure and water uplift. Retaining walls shall be designed for a safety factor of 1.5 against lateral sliding and overturning.

CHANGE SIGNIFICANCE.   There are now new criteria for the design of retaining walls provided in the 2006 code. Where the 2003 code had no specific provisions for retaining walls, and because retaining walls are a normal feature in many residential sites, specific criteria compatible with the IBC have been provided in the 2006 IRC.

This needed section is consistent with requirements found in the IBC and designates when a wall becomes a retaining wall. Further, the proposed change provides specific criteria for which a retaining wall must be designed. Retaining walls are regular features of residential construction and should be addressed in the IRC.

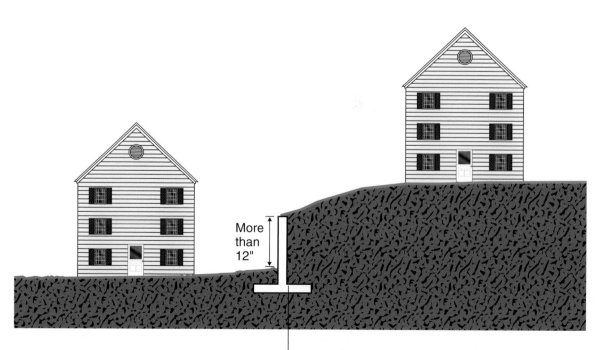

More than 12"

Design retaining wall with a factor of safety of 1.5 against sliding and overturning.

CHANGE TYPE. Modification

CHANGE SUMMARY. This change will make the IRC more consistent with the IBC and requires dampproofing of concrete and masonry foundations for all interior and below-grade spaces and not just habitable and usable spaces.

2006 CODE: R406.1 Concrete and Masonry Foundation Dampproofing. Except where required to be waterproofed by Section R406.2, foundation walls that retain earth and enclose ~~habitable or usable spaces located~~ <u>interior spaces and floors</u> below grade shall be dampproofed from the top of the footing to the finished grade. Masonry walls shall have not less than ⅜ inch (9.5 mm) portland cement parging applied to the exterior of the wall. The parging shall be dampproofed <u>in accordance with one of the following:</u> ~~with a bituminous coating, 3 pounds per square yard (1.63 kg/m²) of acrylic modified cement, ⅛-inch (3.2 mm) coat of surface bonding mortar complying with ASTM C 887 or any material permitted for waterproofing in Section R406.2.~~

R406.1 continues

R406.1

Concrete and Masonry Foundation Dampproofing

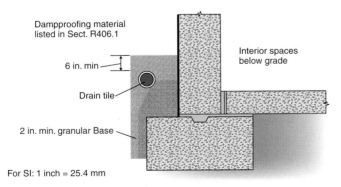

Parging

Bituminous coating or other material listed in Sect. R406.1

Gravel or stone fill

Interior spaces below grade

Drain Tile

Dampproofing of Masonry Foundation Walls

Dampproofing material listed in Sect. R406.1

6 in. min

Drain tile

Interior spaces below grade

2 in. min. granular Base

For SI: 1 inch = 25.4 mm

Dampproofing of Concrete Foundation Walls

R406.1 continued

1. Bituminous coating.
2. 3 pounds per square yard (1.63 kg/m^2) of acrylic modified cement.
3. $\frac{1}{8}$-inch (3.2 mm) coat of surface-bonding cement complying with ASTM C 887.
4. Any material permitted for waterproofing in Section R406.2.
5. Other approved methods or materials.

Exception: Parging of unit masonry walls is not required where a material is approved for direct application to the masonry.

Concrete walls shall be dampproofed by applying any one of the above listed dampproofing materials or any one of the waterproofing materials listed in Section R406.2 to the exterior of the wall.

CHANGE SIGNIFICANCE. This change will make the IRC more consistent with the IBC. The 2003 IRC requires dampproofing of foundation walls only where such walls enclose habitable and usable spaces. As a result, under the 2003 IRC, those foundation walls that enclose crawl spaces and similar areas would remain vulnerable to moisture penetration and dampness. This change intends to provide the same protection that was previously provided for habitable and usable spaces against moisture penetration to all interior spaces, including crawl spaces and subfloors below finish-outside grade. The wording "interior spaces and floors" is consistent with the IBC. Additionally, the new exception would allow dampproofing materials to be applied directly to the masonry substrate without the interim step of a parging application. Such methods for masonry are common and include approved grout coatings, cement-based paints, and/or bituminous coatings. This exception acknowledges that there are such coatings available and being used successfully and eliminates the process of having to go through the "alternative materials, design and methods of construction and equipment" found in Section R104.11.

CHANGE TYPE.   Modification and Addition

CHANGE SUMMARY.   This change will make the IRC more consistent with the IBC and requires waterproofing of concrete and masonry foundations for all interior and below-grade spaces and not just habitable and usable spaces.

2006 CODE: R406.2 Concrete and Masonry Foundation Waterproofing.   In areas where a high water table or other severe soil-water conditions are known to exist, exterior foundation walls that retain earth and enclose ~~habitable or usable space located~~ interior spaces and floors below grade shall be waterproofed ~~with a membrane extending~~ from the top of the footing to the finished grade. ~~The membrane waterproofing shall consist of 2-ply hot-mopped felts, 55-pound (25 kg) roll roofing, 6-mil (0.15 mm) polyvinyl-chloride, 6-mil (0.15 mm) polyethylene or 40-mil (1 mm) polymer-modified asphalt. The joints in the membrane shall be lapped and sealed with an adhesive compatible with the waterproofing membrane.~~ Walls shall be waterproofed in accordance with one of the following:

1. 2-ply hot-mopped felts
2. 55 pound (25 kg) roll roofing
3. 6-mil (0.15 mm) polyvinyl chloride
4. 6-mil (0.15 mm) polyethylene
5. 40-mil (1 mm) polymer-modified asphalt
6. 60-mil (1.5 mm) flexible polymer cement
7. ⅛ inch cement-based, fiber-reinforced, waterproof coating
8. 60-mil (1.5 mm) solvent-free, liquid-applied synthetic rubber

Exception: Organic solvent-based products such as hydrocarbons, chlorinated hydrocarbons, ketones, and esters

R406.2 continues

R406.2

Concrete and Masonry Foundation Waterproofing

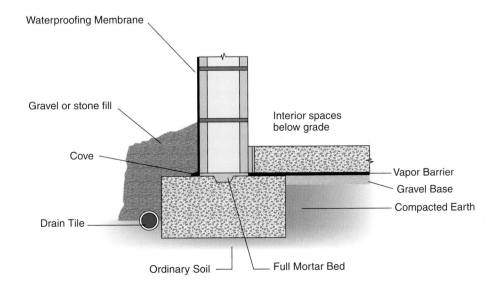

Waterproofing Membrane

Gravel or stone fill

Cove

Drain Tile

Interior spaces below grade

Vapor Barrier

Gravel Base

Compacted Earth

Ordinary Soil

Full Mortar Bed

R406.2 continued

shall not be used for ICF walls with expanded polystyrene form material. Plastic roofing cements, acrylic coatings, latex coatings, mortars, and pargings are permitted to be used to seal ICF walls. Cold setting asphalt or hot asphalt shall conform to type C of ASTM D449. Hot asphalt shall be applied at a temperature of less than 200 degrees.

<u>All joints in membrane waterproofing shall be lapped and sealed with an adhesive compatible with the membrane.</u>

CHANGE SIGNIFICANCE. This change will make the IRC more consistent with the IBC. The 2003 IRC requires waterproofing of foundation walls only where such walls enclose habitable and usable spaces. As a result, under the 2003 IRC, those foundation walls that enclose crawl spaces and similar areas would remain vulnerable to water and moisture penetration. This change intends to provide the same protection that was previously provided for habitable and usable spaces against water and moisture penetration to all interior spaces, including crawl spaces and subfloors below finish-outside grade. The wording "interior spaces and floors" is consistent with the IBC. Additionally, three new prescriptive waterproofing methods are added to the previous five options to avoid having to go through the "alternative materials, design and methods of construction and equipment" found in Section R104.11 and to provide more options. These three new options are the 60-mil flexible polymer cement, the ⅛-inch cement-based, fiber-reinforced, waterproof coating, and the 60-mil solvent-free, liquid-applied synthetic rubber. The new exception is specific to the application of waterproofing methods to insulating concrete forms (ICF) walls only and is included because it is needed for ICF walls.

CHANGE TYPE.   Modification

CHANGE SUMMARY.   The criterion for reduction of ventilation openings has been deleted. The criteria for elimination of ventilation openings for crawl and under-floor spaces have been revised and relocated to new Section 408.3, but a new exception clarifying the location of ventilation openings in the perimeter of buildings has been added.

2006 CODE: R408.2 Openings for Under-Floor Ventilation. The minimum net area of ventilation openings shall not be less than 1 square foot (0.0929 m²) for each 150 square feet (14 m²) of under-floor space area. One such ventilating opening shall be within 3 feet (914 mm) of each corner of the building. Ventilation openings shall be covered for their height and width with any of the following materials provided that the least dimension of the covering shall not exceed ¼ inch (6.4 mm):

1. Perforated sheet metal plates not less than 0.070 inch (1.8 mm) thick.
2. Expanded sheet metal plates not less than 0.047 inch (1.2 mm) thick.
3. Cast iron grills or grating.
4. Extruded load-bearing brick vents.
5. Hardware cloth of 0.035 inch (0.89 mm) wire or heavier.
6. Corrosion-resistant wire mesh, with the least dimension being ⅛ inch (3.2 mm).

~~**Exceptions:**~~
~~1. Where warranted by climatic conditions, ventilation openings to the outdoors are not required if ventilation openings to the interior are provided.~~
~~2. The total area of ventilation openings may be reduced to ¹⁄₁₅₀₀ of the under-floor area where the ground surface is~~

R408.2, R408.3 continues

R408.2, R408.3

Openings for Under-Floor Ventilation (R408.2) and Unvented Crawl Space (R408.3)

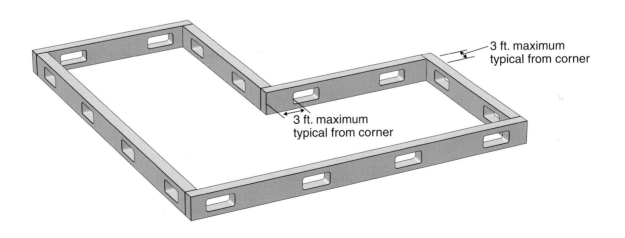

3 ft. maximum typical from corner

3 ft. maximum typical from corner

R408.2, R408.3 continued

~~treated with an approved vapor retarder material and the required openings are placed so as to provide cross-ventilation of the space. The installation of operable louvers shall not be prohibited.~~

~~3. Under-floor spaces used as supply plenums for distribution of heated and cooled air shall comply with the requirements of Section M1601.4~~

~~4. Ventilation openings are not required where continuously operated mechanical ventilation is provided at a rate of 1.0 cfm (10m2) for each 50 square feet (1.02 L/s) of under-floor space floor area and ground surface is covered with an approved vapor retarder material.~~

~~5. Ventilation openings are not required when the ground surface is covered with an approved vapor retarder material, the space is supplied with conditioned air and the perimeter walls are insulated in accordance with Section N1102.1.7.~~

R408.3 Unvented Crawl Space. <u>Ventilation openings in under-floor spaces specified in Sections R408.1 and R408.2 shall not be required where:</u>

1. <u>Exposed earth is covered with a continuous vapor retarder. All joints of the vapor retarder shall overlap by 6 inches (152 mm) and shall be sealed or taped. The edges of the vapor retarder shall extend at least 6 inches (152 mm) up the stem wall and shall be attached and sealed to the stem wall; and</u>

2. <u>One of the following is provided for the under-floor space:</u>
   a. <u>Continuously operated mechanical exhaust ventilation at a rate equal to 1 cfm (0.47 L/s) for each 50 ft^2 (4.7 m^2) of crawlspace floor area, including a return pathway to the common area (such as a duct or transfer grille) and perimeter walls insulated in accordance with Section N1102.2.8 or</u>
   b. <u>Conditioned air supply sized to deliver at a rate equal to 1 cfm (0.47 L/s) for each 50 ft^2 (4.7 m^2) of under-floor area, including a return air pathway to the common area (such as a duct or transfer grille), and perimeter walls insulated in accordance with Section N1102.2.8, or</u>
   c. <u>Plenum complying with M1601.4, if under floor spaces used as a plenum.</u>

CHANGE SIGNIFICANCE. The 2003 code Sections R408.1 and R408.2 establish the requirements for ventilated under-floor spaces and the amount of ventilation to be provided. There are five exceptions in the 2003 code that allow either the elimination or the reduction of ventilation for such spaces. The 2006 code has now eliminated all five exceptions and reformatted them into new Section 408.3 but has added a new exception to clarify the location of ventilation openings. The new Section 408.3, titled "Unvented crawl space," no longer has an option for reduction of ventilation, which was exception 2 under the 2003 code, and only provides for cases where ventilation is allowed to be completely eliminated. This change simplifies the application of the code by eliminating undefined and vague phrases such as "where warranted by climatic conditions" and simplifies the remainder of the provisions within the exceptions. Simplifying the requirements is intended to improve and make the requirements easier to understand and enforce, thereby resulting in more complying buildings, which will therefore improve health and energy conservation. The intent is for this revision to keep the levels of energy performance at those intended by the 2003 code.

R502.2.1 and R602.10.8

Framing at Braced Wall Lines (R502.2.1) and Connections (R602.10.8)

CHANGE TYPE. Addition

CHANGE SUMMARY. Framing requirements for transfer of lateral loads by means of blocking and rim joists have been added for the proper continuity of the load path.

2006 CODE: <u>**R502.2.1 Framing at Braced Wall Lines.** A load path for lateral forces shall be provided between floor framing and braced wall panels located above or below a floor, as specified in Section R602.10.8.</u>

R602.10.8 Connections. Braced wall line sole plates shall be fastened to the floor framing and top plates shall be connected to the framing above in accordance with Table R602.3(1). Sills shall be fastened to the foundation or slab in accordance with Sections R403.1.6 and R602.11. <u>Where joists are perpendicular to the braced wall lines above, blocking shall be provided under and in line with the braced wall panels. Where joists are perpendicular to braced wall lines below, blocking shall be provided over and in line with the braced wall panels. Where joists are parallel to braced wall lines above or below, a rim joist or other parallel framing member shall be provided at the wall to permit fastening per Table R602.3(1).</u>

CHANGE SIGNIFICANCE. The lateral forces induced by wind and earthquake are transferred to the building foundation through the roof and floor systems and the braced wall lines to which these framing members are attached. The braced wall lines in turn transfer such loads to the foundation by the sill plate and its connection to the foundation system. The proper and successful transfer of all loads from the building envelope system to the foundation is accomplished by

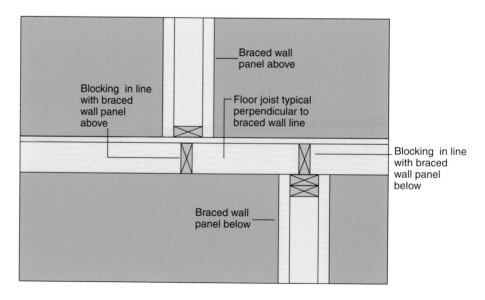

Braced wall panel above

Blocking in line with braced wall panel above

Floor joist typical perpendicular to braced wall line

Blocking in line with braced wall panel below

Braced wall panel below

Setion view

proper creation of a complete load path. Whereas the 2003 IRC lacks specific guidance for creating such a load path in certain situations, the 2006 code inserts new text, both in the chapter on floor construction and the chapter on wall construction to address the conditions that had not been previously addressed. Section R502.2.1 creates the charging provisions for a complete load path, and for the details of such framing it refers the user to Section R602.10.8. The conditions that were not addressed previously but are now addressed in Section R602.10.8 are the framing needed when braced wall panels occur below a floor or when framing is parallel to a braced wall panel. Depending on which of the two conditions is present, the code requires the use of either blocking in line with the braced wall lines or the use of rim joists or other parallel framing members to permit proper fastening required by other sections of the code.

Table R502.5(1)

Girder Spans and Header Spans for Exterior Bearing Walls

CHANGE TYPE. Addition

CHANGE SUMMARY. New girder and header spans have been added for the 70-psf ground snow load, within Table R502.5(1).

2006 CODE: Table R502.5(1) Girder Spans and Header Spans for Exterior Bearing Walls (see pages 95–96)

CHANGE SIGNIFICANCE. The 2003 code Table R502.5(1) contains spans for girders and headers for ground snow loads of 30 and 50 psf only. The prescriptive spans provided for girders and headers cannot be used for many parts of the country where the ground snow load is more than 50 psf. Additionally, the 2003 IRC contains rafter-span tables for ground snow loads of 70 psf. The new spans in the 2006 IRC provide additional prescriptive information for those parts of the country where the ground snow load is greater than 50 psf and eliminate the need for engineering solutions.

TABLE R502.5(1)   Girder Spans[a] and Header Spans[a] for Exterior Bearing Walls

(Maximum spans for Douglas fir-larch, hem-fir, southern pine, and spruce-pine-fir[b] and required number of jack studs)

Girders and Headers Supporting	Size	Ground Snow Load (psf)[e] — Building Width[c] (feet)																	
		30						50						70					
		20		28		36		20		28		36		20		28		36	
		Span	NJ[d]	Span	NJ[d]	Span	NJ[d]	Span	NJ[d]	Span	NJ[d]	Span	NJ[d]	Span	NJ[d]	Span	NJ[d]	Span	NJ[d]
Roof and ceiling	2-2×4	3-6	1	3-2	1	2-10	1	3-2	1	2-9	1	2-6	1	2-10	1	2-6	1	2-3	1
	2-2×6	5-5	1	4-8	1	4-2	1	4-8	1	4-1	1	3-8	2	4-2	1	3-8	2	3-3	2
	2-2×8	6-10	1	5-11	2	5-4	2	5-11	2	5-2	2	4-7	2	5-4	2	4-7	2	4-1	2
	2-2×10	8-5	2	7-3	2	6-6	2	7-3	2	6-3	2	5-7	2	6-6	2	5-7	2	5-0	2
	2-2×12	9-9	2	8-5	2	7-6	2	8-5	2	7-3	2	6-6	2	7-6	2	6-6	2	5-10	3
	3-2×8	8-4	1	7-5	1	6-8	1	7-5	1	6-5	2	5-9	2	6-8	1	5-9	2	5-2	2
	3-2×10	10-6	1	9-1	2	8-2	2	9-1	2	7-10	2	7-0	2	8-2	2	7-0	2	6-4	2
	3-2×12	12-2	2	10-7	2	9-5	2	10-7	2	9-2	2	8-2	2	9-5	2	8-2	2	7-4	2
	4-2×8	9-2	1	8-4	1	7-8	1	8-4	1	7-5	1	6-8	1	7-8	1	6-8	1	5-11	2
	4-2×10	11-8	1	10-6	1	9-5	2	10-6	1	9-1	2	8-2	2	9-5	2	8-2	2	7-3	2
	4-2×12	14-1	1	12-2	2	10-11	2	12-2	1	10-7	2	9-5	2	10-11	2	9-5	2	8-5	2
Roof, ceiling, and one center-bearing floor	2-2×4	3-1	1	2-9	1	2-5	1	2-9	1	2-5	1	2-2	1	2-7	1	2-3	1	2-0	1
	2-2×6	4-6	1	4-0	1	3-7	2	4-1	1	3-7	1	3-3	2	3-9	2	3-3	2	2-11	2
	2-2×8	5-9	2	5-0	2	4-6	2	5-2	2	4-6	2	4-1	2	4-9	2	4-2	2	3-9	2
	2-2×10	7-0	2	6-2	2	5-6	2	6-4	2	5-6	2	5-0	2	5-9	2	5-1	2	4-7	3
	2-2×12	8-1	2	7-1	2	6-5	2	7-4	2	6-5	2	5-9	2	6-8	2	5-10	3	5-3	3
	3-2×8	7-2	1	6-3	2	5-8	2	6-5	2	5-8	2	5-1	2	5-11	2	5-2	2	4-8	2
	3-2×10	8-9	2	7-8	2	6-11	2	7-11	2	6-11	2	6-3	2	7-3	2	6-4	2	5-8	2
	3-2×12	10-2	2	8-11	2	8-0	2	9-2	2	8-0	2	7-3	2	8-5	2	7-4	2	6-7	2
	4-2×8	8-1	1	7-3	1	6-7	1	7-5	1	6-6	1	5-11	1	6-10	1	6-0	1	5-5	2
	4-2×10	10-1	1	8-10	2	8-0	2	9-1	2	8-0	2	7-2	2	8-4	2	7-4	2	6-7	2
	4-2×12	11-9	2	10-3	2	9-3	2	10-7	2	9-3	2	8-4	2	9-8	2	8-6	2	7-7	2

Table R502.5(1) continues

Table R502.5(1) continued

Girder and header size	Span	NJ	Span	NJ	Span	NJ	Span	NJ	Span	NJ	Span	NJ	Span	NJ	Span	NJ	Span	NJ
Roof, ceiling, and one clear span floor																		
2-2×4	2-8	1	2-4	1	2-1	1	2-7	1	2-3	1	2-0	1	2-5	1	2-1	1	1-10	1
2-2×6	3-11	1	3-5	2	3-0	2	3-10	2	3-4	2	3-0	2	3-6	2	3-1	2	2-9	2
2-2×8	5-0	2	4-4	2	3-10	2	4-10	2	4-2	2	3-9	2	4-6	2	3-11	2	3-6	2
2-2×10	6-1	2	5-3	2	4-8	2	5-11	2	5-1	2	4-7	3	5-6	3	4-9	2	4-3	3
2-2×12	7-1	2	6-1	3	5-5	3	6-10	3	5-11	3	5-4	3	6-4	3	5-6	3	5-0	3
3-2×8	6-3	2	5-5	2	4-10	2	6-1	2	5-3	2	4-8	2	5-7	2	4-11	2	4-5	2
3-2×10	7-7	2	6-7	2	5-11	2	7-5	2	6-5	2	5-9	2	6-10	2	6-0	2	5-4	2
3-2×12	8-10	2	7-8	2	6-10	2	8-7	2	7-5	2	6-8	2	7-11	2	6-11	2	6-3	3
4-2×8	7-2	1	6-3	2	5-7	2	7-0	1	6-1	2	5-5	2	6-6	1	5-8	1	5-1	2
4-2×10	8-9	2	7-7	2	6-10	2	8-7	2	7-5	2	6-7	2	7-11	2	6-11	2	6-2	2
4-2×12	10-2	2	8-10	2	7-11	2	9-11	2	8-7	2	7-8	2	9-2	2	8-0	2	7-2	2
Roof, ceiling, and two center-bearing floors																		
2-2×4	2-7	1	2-3	1	2-0	1	2-6	1	2-2	1	1-11	1	2-4	1	2-0	1	1-9	1
2-2×6	3-9	2	3-3	2	2-11	2	3-8	2	3-2	2	2-10	2	3-5	2	3-0	2	2-8	2
2-2×8	4-9	2	4-2	2	3-9	2	4-7	2	4-0	2	3-8	2	4-4	2	3-9	2	3-5	2
2-2×10	5-9	2	5-1	2	4-7	3	5-8	2	4-11	2	4-5	2	5-3	3	4-7	2	4-2	3
2-2×12	6-8	2	5-10	3	5-3	3	6-6	2	5-9	2	5-2	3	6-1	3	5-5	2	4-10	3
3-2×8	5-11	2	5-2	2	4-8	2	5-9	2	5-1	2	4-7	2	5-5	2	4-9	2	4-3	2
3-2×10	7-3	2	6-4	2	5-8	2	7-1	2	6-2	2	5-7	2	6-7	2	5-9	2	5-3	2
3-2×12	8-5	2	7-4	2	6-7	2	8-2	2	7-2	2	6-5	2	7-8	2	6-9	2	6-1	3
4-2×8	6-10	1	6-0	2	5-5	2	6-8	1	5-10	2	5-3	2	6-3	2	5-6	2	4-11	2
4-2×10	8-4	2	7-4	2	6-7	2	8-2	2	7-2	2	6-5	2	7-7	2	6-8	2	6-0	2
4-2×12	9-8	2	8-6	2	7-8	2	9-5	2	8-3	2	7-5	2	8-10	2	7-9	2	7-0	2
Roof, ceiling, and two clear span floors																		
2-2×4	2-1	1	1-8	1	1-6	1	2-0	2	1-8	2	1-5	1	2-0	2	1-8	1	1-5	2
2-2×6	3-1	2	2-8	2	2-4	2	3-0	2	2-7	2	2-3	2	2-11	2	2-7	2	2-3	2
2-2×8	3-10	2	3-4	3	3-0	3	3-10	3	3-4	2	2-11	3	3-9	3	3-3	2	2-11	3
2-2×10	4-9	2	4-1	3	3-8	3	4-8	3	4-0	3	3-7	3	4-7	3	4-0	3	3-6	3
2-2×12	5-6	3	4-9	3	4-3	3	5-5	3	4-8	3	4-2	3	5-4	3	4-7	3	4-1	4
3-2×8	4-10	2	4-2	2	3-9	2	4-9	2	4-1	2	3-8	2	4-8	2	4-1	2	3-8	2
3-2×10	5-11	2	5-1	2	4-7	3	5-10	2	5-0	2	4-6	3	5-9	2	4-11	2	4-5	3
3-2×12	6-10	3	5-11	3	5-4	3	6-9	3	5-10	3	5-3	3	6-8	3	5-9	3	5-2	3
4-2×8	5-7	2	4-10	2	4-4	2	5-6	2	4-9	2	4-3	2	5-5	2	4-8	2	4-2	2
4-2×10	6-10	2	5-11	2	5-3	2	6-9	2	5-10	2	5-2	2	6-7	2	5-9	2	5-1	2
4-2×12	7-11	2	6-10	2	6-2	3	7-9	2	6-9	2	6-0	3	7-8	2	6-8	2	5-11	3

For SI: 1 inch = 25.4 mm, 1 pound per square foot = 0.0479 KPa.

a. Spans are given in feet and inches.
b. Tabulated values assume #2 grade lumber.
c. Building width is measured perpendicular to the ridge. For widths between those shown, spans are permitted to be interpolated.
d. NJ = Number of jack studs required to support each end. Where the number of required jack studs equals one, the header is permitted to be supported by an approved framing anchor attached to the full-height wall stud and to the header.
e. Use 30 psf ground snow load for cases in which ground snow load is less than 30 psf and the roof live load is equal to or less than 20 psf.

CHANGE TYPE. Clarification

CHANGE SUMMARY. The reference to wood-frame floor and floor-ceiling assemblies has been deleted to clarify that there are no fireblocking requirements for floor and floor-ceiling assemblies.

2006 CODE: R502.13 Fireblocking Required. Fireblocking shall be provided ~~in wood-frame floor construction and floor-ceiling assemblies~~ in accordance with Section R602.8.

CHANGE SIGNIFICANCE. The 2003 IRC text requires fireblocking for wood-frame floor and floor-ceiling assemblies on the basis of details of Section 602.8. In Section 602.8 there are no criteria for fireblocking for floor and floor-ceiling assemblies. The code never intended to require fireblocking in floor and floor-ceiling assemblies; however, the reference to Section 602.8 led many code users to interpret the code to mean that the criteria outlined in Section 602.8 for walls also applies to floor and floor-ceiling assemblies. The phrase that has been deleted implied that there were such fireblocking requirements for floor and floor-ceiling assemblies. The 2006 IRC language in this section is intended to clarify that even though there are fireblocking requirements in the code, the application is based on the scoping in Section 602.8 and applies to walls and conditions specifically designated in that section.

R502.13
Fireblocking Required

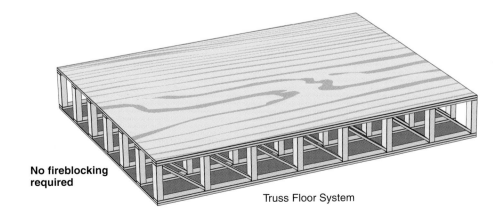

No fireblocking required

Truss Floor System

Table R503.2.1.1(1)

Allowable Spans and Loads for Wood Structural Panels for Roof and Subfloor Sheathing and Combination Subfloor Underlayment

CHANGE TYPE. Addition

CHANGE SUMMARY. A new column entry has been added to include the most common framing spacings of 16 inches and 24 inches and larger live loads than what is found in the 2003 code table. For clarification purposes, the reference to APA (Engineered Wood Association) E30, *APA Engineered Wood Construction Guide,* has also been inserted into Sections R503.2.2 and R803.2.2.

2006 CODE: Table R503.2.1.1(1) Allowable Spans and Loads for Wood Structural Panels for Roof and Subfloor Sheathing and Combination Subfloor Underlayment (see page 99)

CHANGE SIGNIFICANCE. The 2003 IRC Table R503.2.1.1 (1) has two flaws in the prescriptive application of the code. One is that the highest loads for which span information is provided are 40 psf for live and 50 psf for total load, and the other is that a direct link between framing spacings of 16 inches and 24 inches—the most commonly used in the industry—and the tabulated loads is not always easy and straightforward. The 2006 IRC has addressed both of these issues by inserting new information into Table R503.2.1.1 (1). This new information is taken from the APA E30 standard, *APA Engineered Wood Construction Guide.* The new column provides allowable loading values for all thicknesses of roof and subfloor sheathing, at standardized common-practice spacing of structural members.

The new column also provides information for allowable live loads of up to 305 psf for sheathing and 290 psf for underlayment. The new higher live loads within the table are intended to provide the code user prescriptive information for locations with higher

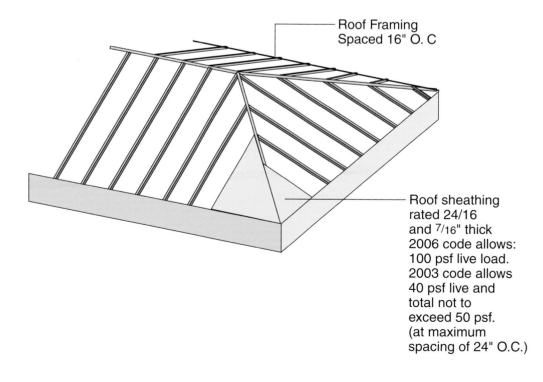

Roof Framing
Spaced 16" O. C

Roof sheathing
rated 24/16
and 7/16" thick
2006 code allows:
100 psf live load.
2003 code allows
40 psf live and
total not to
exceed 50 psf.
(at maximum
spacing of 24" O.C.)

TABLE R503.2.1.1(1)   Allowable Spans and Loads for Wood Structural Panels for Roof and Subfloor Sheathing and Combination Subfloor Underlayment[a,b,c]

Span Rating	Minimum Nominal Panel Thickness (inch)	Allowable Live Load (psf)[h,l]		Maximum Span (inches)		Load (pounds per square foot, at maximum span)		Maximum Span (inches)
		Span @ 16" o.c.	Span @ 24" o.c.	With Edge Support[d]	Without Edge Support	Total Load	Live Load	
Sheathing[e]				Roof[f]				Subfloor[j]
12/0	$5/16$	—	—	12	12	40	30	0
16/0	$5/16$	30	—	16	16	40	30	0
20/0	$5/16$	50	—	20	20	40	30	0
24/0	$3/8$	100	30	24	20[g]	40	30	0
24/16	$7/16$	100	40	24	24	50	40	16
32/16	$15/32, 1/2$	180	70	32	28	40	30	16[h]
40/20	$19/32, 5/8$	305	30	40	32	40	30	20[h,i]
48/24	$23/32, 3/4$	—	175	48	36	45	35	24
60/32	$7/8$	—	305	60	48	45	35	32
Underlayment, C-C plugged, single floor[e]				Roof[f]				**Combination Subfloor Underlayment[k]**
16 o.c.	$19/32, 5/8$	100	40	24	24	50	40	16[i]
20 o.c.	$19/32, 5/8$	150	60	32	32	40	30	20[i,j]
24 o.c.	$23/32, 3/4$	240	100	48	36	35	25	24
32 o.c.	$7/8$	—	185	48	40	50	40	32
48 o.c.	$13/32, 1 1/8$	—	290	60	48	50	40	48

For SI: 1 inch = 25.4 mm. 1 pound per square foot = 0.0479 KPa.

a. The allowable total loads were determined using a dead load of 10 psf. If the dead load exceeds 10 psf, then the live load shall be reduced accordingly.

b. Panels continuous over two or more spans with long dimension perpendicular to supports. Spans shall be limited to values shown because of possible effect of concentrated loads.

c. Applies to panels 24 inches or wider.

d. Lumber blocking, panel edge clips (one midway between each support, except two equally spaced between supports when span is 48 inches), tongue-and-groove panel edges, or other approved type of edge support.

e. Includes Structural 1 panels in these grades.

f. Uniform load deflection limitation: $1/180$ of span under live load plus dead load, $1/240$ of span under live load only.

g. Maximum span 24 inches for $15/32$- and $1/2$-inch panels.

h. Maximum span 24 inches where $3/4$-inch wood finish flooring is installed at right angles to joists.

i. Maximum span 24 inches where 1.5 inches of lightweight concrete or approved cellular concrete is placed over the subfloor.

j. Unsupported edges shall have tongue-and-groove joints or shall be supported with blocking unless minimum nominal $1/4$-inch thick underlayment with end and edge joints offset at least 2 inches or 1.5 inches of lightweight concrete or approved cellular concrete is placed over the subfloor, or $3/4$-inch wood finish flooring is installed at right angles to the supports. Allowable uniform live load at maximum span, based on deflection of $1/360$ of span, is 100 psf.

k. Unsupported edges shall have tongue-and-groove joints or shall be supported by blocking unless nominal $1/4$-inch-thick underlayment with end and edge joints offset at least 2 inches or $3/4$-inch wood finish flooring is installed at right angles to the supports. Allowable uniform live load at maximum span, based on deflection of $1/360$ of span, is 100 psf, except panels with a span rating of 48 on center are limited to 65 psf total uniform load at maximum span.

l. Allowable live load values at spans of 16" o.c. and 24" o.c. taken from reference standard APA-E30, APA Engineered Wood Construction Guide. Refer to reference standard for allowable spans not listed in the table.

Table R503.2.1.1(1) continues

Table R503.2.1.1(1) continued

snow loads, which would include areas with ground snow loads up to 70 psf. It should be noted, however, that the new column heading "Allowable Live Load" is technically inconsistent for use with snow loads, because the definition of live load in Section R202 is "those loads produced by the use and occupancy of the building or other structure and do not include construction or environmental loads such as wind load, snow load, rain load, earthquake load, flood load, or dead load."

CHANGE TYPE.   Clarification

CHANGE SUMMARY.   The 10-foot story limitation for the applicability of steel floor framing has been deleted to clarify that provisions contained in this section apply where the stud height is 10 feet and the floor framing height does not exceed 16 inches. The maximum width limitation of 36 feet has also been increased to 40 feet.

2006 CODE: R505.1.1 Applicability Limits.   The provisions of this section shall control the construction of steel floor framing for buildings not greater than 60 feet (18,288 mm) in length perpendicular to the joist span, not greater than ~~36 feet (10,973 mm)~~ 40 feet (12,192 mm) in width parallel to the joist span, and not greater than two stories in height ~~with each story not greater than 10 feet (3,048 mm) high~~.   Steel floor framing constructed in accordance with the provisions of this section shall be limited to sites subjected to a maximum design wind speed of 110 miles per hour (49 m/s) Exposure A, B, or C and a maximum ground snow load of 70 pounds per square foot (3.35 KPa).

CHANGE SIGNIFICANCE.   As written in the 2003 IRC, there is a conflict for the limitation imposed on steel floor framing provisions.

R505.1.1 continues

R505.1.1

Applicability Limits for Steel Floor Framing

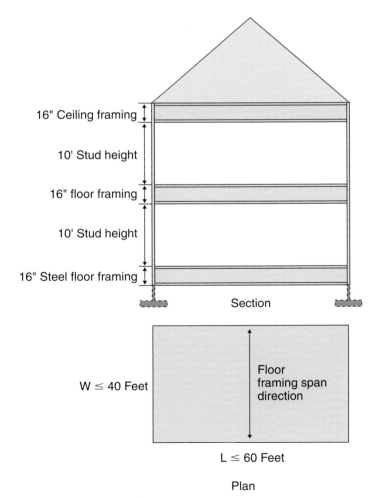

16" Ceiling framing

10' Stud height

16" floor framing

10' Stud height

16" Steel floor framing

Section

W ≤ 40 Feet

Floor framing span direction

L ≤ 60 Feet

Plan

R505.1.1 continued   Whereas Section R301.3 item 2 allows the limitation to be a stud height of 10 feet plus floor framing depth, not to exceed 16 inches, Section R505.1.1 sets the limitation at a story height of 10 feet. Story is defined in Section R202 as "that portion of a building included between the upper surface of a floor and the upper surface of the floor or roof next above," and therefore the application of Section R505.1.1 appeared to be limited to 10 feet for the entire story of stud height and floor or roof assembly. This same clarification has been made in Sections R603.1.1 and R804.1.1 in the same manner. The maximum 36 feet limitation for building width has also been changed from 36 feet to 40 feet for better coordination between the IRC and the AISI standard ANSI/COFS/PM-2001 and PM supplement-2004 (standard for cold-formed steel framing-prescriptive method for one- and two-family dwellings). These changes should result in a better coordination between the IRC cold-formed-steel framing sections (e.g., floors, walls, and roof), and the design criteria provisions in Chapter 3, and the AISI standards.

R505.1.3 and R505.3.2

Floor Trusses (R505.1.3) and Allowable Joist Spans (R505.3.2)

CHANGE TYPE.   Addition

CHANGE SUMMARY.   A new standard for cold-formed-steel floor truss design, quality assurance, installation, and testing is provided. New blocking provisions and new tables for the span of cold-formed-steel floor joists have been created.

2006 CODE: R505.1.3 Floor Trusses.   The design, quality assurance, installation, and testing of cold-formed steel trusses shall be in accordance with the AISI Standard for Cold-formed Steel Framing-Truss Design (COFS/Truss). Truss members shall not be notched, cut, or altered in any manner without an approved design.

R505.3.2 Allowable Joist Spans.   The clear span of cold-formed steel floor joists shall not exceed the limits set forth in Tables R505.3.2(1), R505.3.2(2), and R505.3.2(3). Floor joists shall have a minimum bearing length of 1.5 inches (38 mm). When continuous joists are used the interior bearing supports shall be located within 2 feet (610 mm) of mid span of the steel joists, and the individual spans shall not exceed the span in Tables R505.3.2(2) and R505.3.2(3). Bearing stiffeners shall be installed at each bearing location in accordance with Section R505.3.4 and as shown in figure R505.3.

Blocking is not required for continuous back-to-back floor joists at bearing supports. Blocking shall be installed between the joists for single continuous floor joists across bearing supports. Blocking shall be spaced at a maximum of 12 feet (3660 mm) on center. Blocking shall consist of C shape or track section with a minimum thickness of 33 mils (0.84 mm). Blocking shall be fastened to each adjacent joist through a 33 mill (0.84 mm) clip angle, bent web of blocking, or

R505.1.3 and R505.3.2 continues

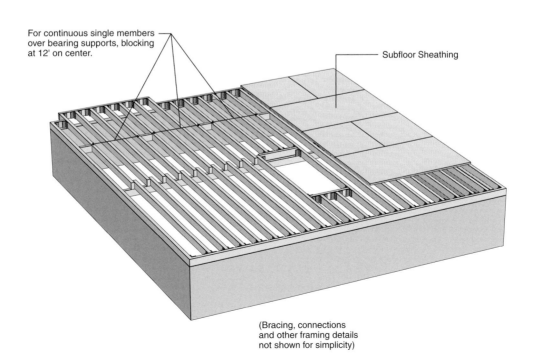

For continuous single members over bearing supports, blocking at 12' on center.

Subfloor Sheathing

(Bracing, connections and other framing details not shown for simplicity)

R505.1.3 and R505.3.2 continued flanges of web stiffeners with two No. 8 screws on each side. The minimum depth of the blocking shall be equal to the depth of the joist minus 2 inches (51 mm). The minimum length of the angle shall be equal to the depth of the joist minus 2 inches (51 mm).

TABLE R505.3.2(1) Allowable Spans for Cold-Formed Steel Joists-Single Spans[a,b] 33 ksi Steel

Joist Designation	30 psf Live Load Spacing (inches)				40 psf Live Load Spacing (inches)			
	12	16	19.2	24	12	16	19.2	24
550S162-33	11'-7"	10'-7"	9'-6"	8'-6"	10'-7"	9'-3"	8'-6"	7'-6"
550S162-43	12'-8"	11'-6"	10'-10"	10'-2"	11'-6"	10'-5"	9'-10"	9'-1"
550S162-54	13'-7"	12'-4"	11'-7"	10'-9"	12'-4"	11'-2"	10'-6"	9'-9"
550S162-68	14'-7"	13'-3"	12'-6"	11'-7"	13'-3"	12'-0"	11'-4"	10'-6"
550S162-97	16'-2"	14'-9"	13'-10"	12'-10"	14'-9"	13'-4"	12'-7"	11'-8"
800S162-33	15'-8"	13'-11"	12'-9"	11'-5"	14'-3"	12'-5"	11'-3"	9'-0"
800S162-43	17'-1"	15'-6"	14'-7"	13'-7"	15'-6"	14'-1"	13'-3"	12'-4"
800S162-54	18'-4"	16'-8"	15'-8"	14'-7"	16'-8"	15'-2"	14'-3"	13'-3"
800S162-68	19'-9"	17'-11"	16'-10"	15'-8"	17'-11"	16'-3"	15'-4"	14'-2"
800S162-97	22'-0"	20'-0"	18'-10"	17'-5"	20'-0"	18'-2"	17'-1"	15'-10"
1000S162-43	20'-6"	18'-8"	17'-6"	15'-8"	18'-8"	16'-11"	15'-6"	13'-11"
1000S162-54	22'-1"	20'-0"	18'-10"	17'-6"	20'-0"	18'-2"	17'-2"	15'-11"
1000S162-68	23'-9"	21'-7"	20'-3"	18'-10"	21'-7"	19'-7"	18'-5"	17'-1"
1000S162-97	26'-6"	24'-1"	22'-8"	21'-0"	24'-1"	21'-10"	20'-7"	19'-1"
1200S162-43	23'-9"	20'-10"	19'-0"	16'-8"	21'-5"	18'-6"	16'-6"	13'-2"
1200S162-54	25'-9"	23'-4"	22'-0"	20'-1"	23'-4"	21'-3"	20'-0"	17'-10"
1200S162-68	27'-8"	25'-1"	23'-8"	21'-11"	25'-1"	22'-10"	21'-6"	21'-1"
1200S162-97	30'-11"	28'-1"	26'-5"	24'-6"	28'-1"	25'-6"	24'-0"	22'-3"

For SI: 1 inch = 25.4 mm. 1 foot = 304.8 mm. 1 pound per square foot = 0.0479 kN/m².

a. Deflection criteria: L/480 for live loads, L/240 for total loads.
b. Floor dead load = 10 psf.

(This is only one of the three new tables found in the 2006 code.)

CHANGE SIGNIFICANCE. The new Section R505.1.3 is included to recognize the truss design standard (ANSI/COFS/Truss Standard for Cold-Formed-Steel Framing-Truss Design). This standard is currently referenced in Section R804, and it is only appropriate that it be included in the floor-framing section to clarify that cold-formed-steel trusses can be designed by professionals with use of this standard. Section R505.3.2 and its associated tables have been revised, and new text and tables have been added primarily to coordinate the IRC with the American Iron and Steel Institute (AISI) Standard ANSI/COFS/PM-2001 and Prescriptive Method (PM) Supplement-2004 (Standard for Cold-Formed Steel Framing—Prescriptive Method for One- and Two-Family Dwellings). The new paragraph on blocking provides detailed provisions for situations where blocking is not required and for cases where blocking is required and then identifies the details of such blocking. The new tables include single-span as well as multiple-span floor-framing members and also recognize the 97-mil-thick members previously not included.

CHANGE TYPE.   Addition

CHANGE SUMMARY.   The placement and support requirements for reinforcement in concrete slab floors on ground has been added.

2006 CODE: **R506.2.4 Reinforcement Support.** Where provided in slabs on ground, reinforcement shall be supported to remain in place from the center to upper one third of the slab, for the duration of the concrete placement.

CHANGE SIGNIFICANCE.   The 2003 IRC Section R506, Concrete Floors (On Ground), contains no provisions for the placement of steel reinforcement, but it does contain provisions for fill, base, and vapor retarder. One of the major problems in slab-on-ground construction is the placement of reinforcement—welded wire mesh or other reinforcing steel—prior to and during the placement of concrete. Often the reinforcement is walked on and pushed to the bottom, rendering the reinforcement completely useless. This new section requires that if reinforcement is used in slab-on-ground construction, it must be placed at a location between the center and the upper one-third of the slab during the entire concrete placement operation. The new section also requires that reinforcement be "supported" to remain in place; hence, it clearly prohibits a field practice whereby the workers lift the reinforcement that has been placed at the bottom as the concrete is being placed. The only time the center to upper-one-third reinforcement location could be deviated from is when the slab on ground is designed by a registered design professional and the design calls for a different reinforcement location.

R506.2.4
Reinforcement Support

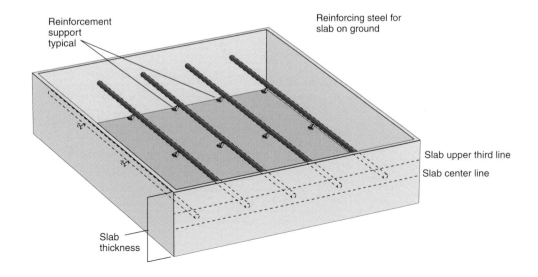

Table R602.3(1)

Fastener Schedule for Structural Members

CHANGE TYPE. Modification

CHANGE SUMMARY. The diameter and length of each nail have been inserted into the table, and a new entry for collar tie to rafter has been inserted and eliminates framing or blocking at roof-plane perimeters expressly for nailing purposes.

2006 CODE:

TABLE R602.3(1)   Fastener Schedule for Structural Members

Description of Building Elements	Number and Type of Fasteners[a,b,c,d]	Spacing of Fasteners
Joist to sill or girder, toe nail	3-8d ($2\frac{1}{2}$″ × 0.113″)	—
1″ × 6″ subfloor or less to each joist, face nail	2-8d ($2\frac{1}{2}$″ × 0.113″) 2 staples, 1 $\frac{3}{4}$″	—
2″ subfloor to joist or girder, blind and face nail	2-16d ($3\frac{1}{2}$″ × 0.135″)	—
Sole plate to joist Subfloor or blocking, face nail	16d ($3\frac{1}{2}$″ × 0.135″)	16″ o.c.
Top or sole plate to stud, end nail	2-16d ($3\frac{1}{2}$″ × 0.135″)	—
Stud to sole plate, toe nail	3-8d ($2\frac{1}{2}$″ × 0.113″) or 2-16d ($3\frac{1}{2}$″ × 0.135″)	—
Double studs, face nail	10d (3″ × 0.128″)	24″ o.c.
Double top plates, face nail	10d (3″ × 0.128″)	24″ o.c.
Sole plate to joist or blocking at braced wall panels	3-16d ($3\frac{1}{2}$″ × 0.135″)	16″ o.c.
Double top plates, minimum 24-inch offset of end joints, face nail in lapped area	8-16d ($3\frac{1}{2}$″ × 0.135″)	—
Blocking between joists or rafters to top plate, toe nail	3-8d ($2\frac{1}{2}$″ × 0.113″)	—
Rim joist to top plate, toe nail	8d ($2\frac{1}{2}$″ × 0.113″)	6″ o.c.
Top plates, laps at corners and intersections, face nail	2-10d (3″ × 0.128″)	—
Built-up header, two pieces with $\frac{1}{2}$″ spacer	16d ($3\frac{1}{2}$″ × 0.135″)	16″ o.c. along each edge
Continued header, two pieces	16d ($3\frac{1}{2}$″ × 0.135″)	16″ o.c. along each edge
Ceiling joists to plate, toe nail	3-8d ($2\frac{1}{2}$″ × 0.113″)	—
Continuous header to stud, toe nail	4-8d ($2\frac{1}{2}$″ × 0.113″)	—
Ceiling joist, laps over partitions, face nail	3-10d (3″ × 0.128″)	—
Ceiling joist to parallel rafters, face nail	3-10d (3″ × 0.128″)	—
Rafter to plate, toe nail	2-16d ($3\frac{1}{2}$″ × 0.135″)	—
1″ brace to each stud and plate, face nail	2-8d ($2\frac{1}{2}$″ × 0.113″) 2 staples, 1 $\frac{3}{4}$″	—
1″ × 6″ sheathing to each bearing, face nail	2-8d ($2\frac{1}{2}$″ × 0.113″) 2 staples, 1 $\frac{3}{4}$″	—
1″ × 8″ sheathing to each bearing, face nail	2-8d ($2\frac{1}{2}$″ × 0.113″) 3 staples, 1 $\frac{3}{4}$″	—
Wider than 1″ × 8″ sheathing to each bearing, face nail	3-8d ($2\frac{1}{2}$″ × 0.113″) 4 staples, 1 $\frac{3}{4}$″	—
Built-up corner studs	10d (3″ × 0.128″)	24″ o.c.

Built-up girders and beams, 2-inch lumber layers	10d (3" × 0.128")	Nail each layer as follows: 32" o.c. at top and bottom and staggered. Two nails at ends and at each splice.
2" planks	2-16d (3½" × 0.135")	At each bearing
Roof rafters to ridge, valley or hip rafters: toe nail	4-16d (3½" × 0.135")	—
face nail	3-16d (3½" × 0.135")	—
Rafter ties to rafters, face	3-8d (2½" × 0.113")	—
Collar tie to rafter, face nail, or 1¼" × 20 gage ridge strap	3-10d (3" × 0.128")	—

Wood structural panels, subfloor, roof and wall sheathing to framing, and particleboard wall sheathing to framing

5/16"-1/2"	6d common (2" × 0.113") nail (subfloor, wall) 8d common (2½" × 0.131") nail (roof)[i]	6	12[g]
19/32"-1"	8d common nail (2½" × 0.131")	6	12[g]
1⅛"-1¼"	10d common (3" × 0.148") nail or 8d (2½" × 0.131") deformed nail	6	12

Description of Building Materials	Description of Fastener[b,c,d,e]	Spacing of Fasteners	
		Edges (inches)[i]	Intermediate Supports[c,e] (inches)

Other wall sheathing[h]

½" regular cellulosic fiberboard sheathing	1½" galvanized roofing nail 6d common (2" × 0.113") nail staple 16 ga. 1 ½ long	3	6
½" structural cellulosic fiberboard sheathing	1½" galvanized roofing nail 8d common (2½" × 0.131") nail staple 16 ga. 1 ½ long	3	6
25/32" structural cellulosic fiberboard sheathing	1¾" galvanized roofing nail 8d common (2½" × 0.131") nail staple 16 ga. 1¾ long	3	6
½" gypsum sheathing	1½" galvanized roofing nail; 6d common (2" × 0.113") nail; staple galvanized. 1½" long: 1¼" screws. Type W or S	4	8
⅝" gypsum sheathing	1¾" galvanized roofing nail; 8d common (2½" × 0.131") nail; staple galvanized. 1⅝ long: 1⅝" screws. Type W or S.	4	8

Wood structural panels, combination subfloor underlayment to framing

¾" and less	6d deformed (2" × 0.120") nail or 8d common (2½" × 0.131")	6	12

Table R602.3(1) continues

Table R602.3(1) continued

TABLE R602.3(1) Fastener Schedule for Structural Members

Description of Building Elements		Number and Type of Fasteners[a,b,c,d]	Spacing of Fasteners
$\frac{7}{8}$"-1"	8d common ($2\frac{1}{2}$" $\times$ 0.131") nail or 8d deformed ($2\frac{1}{2}$" $\times$ 0.120") nail	6	12
$1\frac{1}{8}$"-$1\frac{1}{4}$"	10d common (3" $\times$ 0.148") nail or 8d deformed ($2\frac{1}{2}$" $\times$ 0.120") nail	6	12

For SI: 1 inch = 25.4mm, 1 foot = 304.8 mm, 1 mile per hour = 1.609 km/h.

a. All nails are smooth-common, box or deformed shanks except where otherwise stated. Nails used for framing and sheathing connections shall have minimum average bending yield strengths as shown: 80 ksi (551MPa) for shank diameter of 0.192 inch (20d common nail),90 ksi (620MPa) for shank diameters larger than 0.142 inch but not larger than 0.177 inch, and 100 ksi (689 MPa) for shank diameters of 0.142 inch or less.
b. Staples are 16 gage wire and have a minimum $\frac{7}{16}$-inch on diameter crown width.
c. Nails shall be spaced at not more than 6 inches on center at all supports where spans are 48 inches or greater.
d. Four-foot-by-8-foot or 4-foot-by-9-foot panels shall be applied vertically.
e. Spacing of fasteners not included in this table shall be based on Table R602.3(~~1~~ 2).
f. For regions having basic wind speed of 110 mph or greater, 8d deformed nails shall be used for attaching plywood and wood structural panel roof sheathing to framing within minimum 48-inch distance from gable end walls, if mean roof height is more than 25 feet, up to 35 feet maximum.
g. For regions having basic wind speed of 100 mph or less, nails for attaching wood structural panel roof sheathing to gable end wall framing shall be spaced 6 inches on center. When basic wind speed is greater than 100 mph, nails for attaching panel roof sheathing to intermediate supports shall be spaced 6 inches on center for minimum 48-inch distance from ridges, eaves, and gable end walls; and 4 inches on center to gable end wall framing.
h. Gypsum sheathing shall conform to ASTM C 79 and shall be installed in accordance with GA 253. Fiberboard sheathing shall conform to either AHA 194.1 or ASTM C 208.
i. Spacing of fasteners on floor sheathing panel edges applies to panel edges supported by framing members and required blocking and at all floor perimeters only. Spacing of fasteners on roof sheathing panel edges applies to panel edges supported by framing members and required blocking ~~and at all roof plane perimeters~~. Blocking of roof or floor sheathing panel edges perpendicular to framing members need not be provided except as required by other provisions of this code ~~shall not be required except at intersection of adjacent roof planes~~. Floor ~~and roof~~ perimeter shall be supported by framing members or solid blocking.

CHANGE SIGNIFICANCE. The several changes within this table include insertion of information on nail diameter and length for clarification purposes, the addition of a new entry for collar ties to rafters, and changes to footnote "i" that eliminate the roof-plane perimeter framing and blocking for the express purpose of nailing.

1. Wrong nail sizes are sometimes used in building construction because the pennyweight system used to describe nails in fastening schedules is not universally understood. The focus on pennyweight (8d = 8 penny, 16d = 16 penny, etc.) and not style (common, box, cooler, sinker, finish, etc.) creates problems such as the substitution of box nails for common nails of the same pennyweight. There can be a significant difference between the strength of wood-to-wood connections nailed with nails of the same pennyweight but different styles, due to different diameters; this results in different withdrawal resistance.

2. Minimum prescriptive requirements in Table R602.3(1) for collar ties and ridge straps have been added and are based on minimum connection requirements in the 2001 *Wood Frame Construction Manual* for slopes greater than 3:12, at a wind speed of 100 mph or less, for a roof span of 36 feet or less.

3. The changes in footnote "i" eliminate the requirement to provide framing and blocking at roof plane perimeters for the express purpose of providing roof-sheathing-edge nailing. Fastening at required blocking is emphasized instead. Blocking is currently required by Sections R502.7 and R802.8. Edge fastening of floor and roof diaphragms to blocking at supporting/bracing walls is an important part of maintaining the load path for resistance to wind and seismic forces.

0.113" diameter

2.5"
8 Penny Nail (8d)

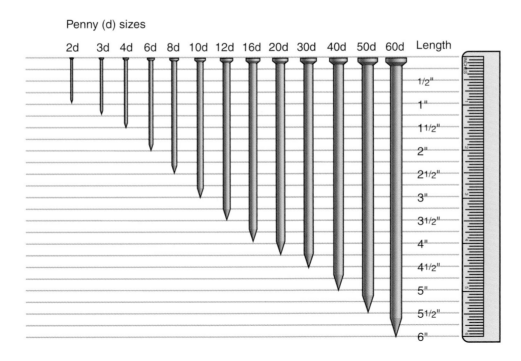

Penny (d) sizes

Table R602.3(2)
Alternate Attachments

CHANGE TYPE. Modification

CHANGE SUMMARY. The table has been revised to increase the length and decrease the spacing of some fasteners used as alternative attachments.

2006 CODE:

TABLE R602.3(2)   Alternate Attachments

Nominal Material Thickness (inches)	Description[a,b] of Fastener and Length (inches)	Spacing[c] of Fasteners	
		Edges (inches)	Intermediate Supports (inches)
Wood structural panels subfloor, roof and wall sheathing to framing and particleboard wall sheathing to framing[f]			
⁵⁄₁₆	~~0.097-0.099 Nail 1¹⁄₂~~	6	~~12~~
	~~Staple 15 ga. 1³⁄₈~~		
	~~Staple 16 ga. 1³⁄₄~~		
³⁄₈	~~Staple 15 ga. 1³⁄₈~~	6	~~12~~
	~~0.097-0.099 Nail 1¹⁄₂~~	4	~~10~~
	~~Staple 16 ga. 1³⁄₄~~	6	~~12~~
~~¹⁵⁄₃₂ and~~ up to ¹⁄₂	Staple 15 ga. ~~1¹⁄₂~~ 1³⁄₄	~~6~~ 4	~~12~~ 8
	0.097-0.099 Nail ~~1⁵⁄₈~~ 2¹⁄₄	3	6
	Staple 16 ga. 1³⁄₄	~~6~~ 3	~~12~~ 6
¹⁹⁄₃₂ and ⁵⁄₈	0.113 Nail ~~1⁷⁄₈~~ 2	3	6
	Staple 15 and 16 ga. ~~1⁵⁄₈~~ 2	~~6~~ 4	~~12~~ 8
	0.097-0.099 Nail ~~1³⁄₄~~ 2¹⁄₄	~~3~~ 4	~~6~~ 8
²³⁄₃₂ and ³⁄₄	Staple 14 ga. ~~1³⁄₄~~ 2	~~6~~ 4	~~12~~ 8
	Staple 15 ga. 1³⁄₄	~~5~~ 3	~~10~~ 6
	0.097-0.099 Nail ~~1⁷⁄₈~~ 2¹⁄₄	~~3~~ 4	~~6~~ 8
	Staple 16 ga. 2	4	8
1	Staple 14 ga. ~~2~~ 2¹⁄₄	~~5~~ 4	~~10~~ 8
	0.113 Nail 2¹⁄₄~~,~~	3	6
	Staple 15 ga. ~~2~~ 2¹⁄₄	4	8
	0.097-0.099 Nail ~~2¹⁄₈~~ 2¹⁄₂	~~3~~ 4	~~6~~ 8

Nominal Material Thickness (inches)	Description[a,b] of Fastener and Length (inches)	Spacing[c] of Fasteners	
		Edges (inches)	Body of panel[d] (inches)
Floor underlayment: plywood-hard board-particleboard[f]			
Plywood			
¹⁄₄ and ⁵⁄₁₆	1¹⁄₄ ring or screw shank nail—minimum 12¹⁄₂ ga. (0.099″) shank diameter	3	6
	Staple 18 ga., ⁷⁄₈, ³⁄₁₆ crown width	2	5
¹¹⁄₃₂, ³⁄₈, ¹⁵⁄₃₂, ~~and~~ ¹⁄₂, and ¹⁹⁄₃₂	1¹⁄₄ ring or screw shank nail—minimum 12¹⁄₂ ga. (0.099) shank diameter	6	8[e]
~~¹⁹⁄₃₂,~~ ⁵⁄₈, ²³⁄₃₂, ³⁄₄	1¹⁄₂ ring or screw shank nail—minimum 12¹⁄₂ ga. (0.099) shank diameter	6	~~12~~ 8
	Staple 16 ga. ~~1¹⁄₄~~ 1¹⁄₂	6	8
Hardboard[f]			
0.200	1¹⁄₂ long ring-grooved underlayment nail	6	6
	4d cement-coated sinker nail	6	6
	Staple 18 ga., ⁷⁄₈ long (plastic coated)	3	6

Particleboard			
¼	4d ring-grooved underlayment nail	3	6
	Staple 18 ga., ⅞ long, ³⁄₁₆ crown	3	6
⅜	6d ring-grooved underlayment nail	6	10
	Staple 16 ga., 1⅛ long, ⅜ crown	3	6
½,⅝	6d ring-grooved underlayment nail	6	10
	Staple 16 ga., 1⅝ long, ⅜ crown	3	6

For SI: 1 inch = 25.4 mm.

a. Nail is a general description and may be T-head, modified round head, or round head.
b. Staples shall have a minimum crown width of ⁷⁄₁₆-inch on diameter except as noted.
c. Nails or staples shall be spaced at not more than 6 inches on center at all supports where spans are 48 inches or greater. Nails or staples shall be spaced at not more than 12 inches on center at intermediate supports for floors.
d. Fasteners shall be placed in a grid pattern throughout the body of the panel.
e. For 5-ply panels, intermediate nails shall be spaced not more than 12 inches on center each way.
f. Hardboard underlayment shall conform to ANSI/AHA A135.4.

CHANGE SIGNIFICANCE. Table R602.3(2) provides alternative fastenings for various materials on the basis of nominal thickness of such materials. As an alternate, it is evident this table must provide fastener lengths, diameters, and spacings in such a manner as to provide the same shear or withdrawal strength as the main Table R602.3(1). The revised table has tried to achieve this equivalency by decreasing the spacing of some fasteners or increasing the length of the others or both. In some cases thinner and shorter fasteners than those listed in Table R602.3(1) are allowed by a reduction in the spacing requirements. The new fastener and fastener spacings are taken from the International Code Council (ICC) Evaluation Service (ES) Legacy Evaluation Report NER-272, revised 1-1-04.

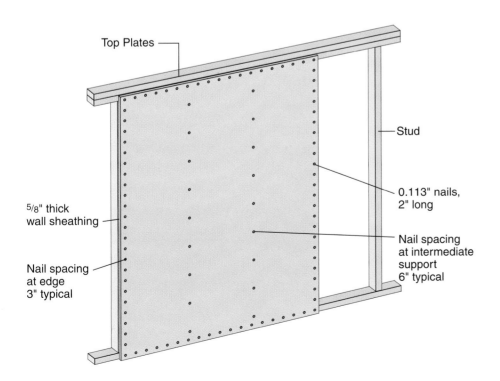

Top Plates

Stud

0.113" nails, 2" long

Nail spacing at intermediate support 6" typical

5/8" thick wall sheathing

Nail spacing at edge 3" typical

R602.3.2

Top Plate

CHANGE TYPE. Clarification

CHANGE SUMMARY. The code text is revised to clarify that joints in top plates of stud walls in conventional construction practices do not need to occur over studs.

2006 CODE: R602.3.2 Top Plate. Wood stud walls shall be capped with a double top plate installed to provide overlapping at corners and intersections with bearing partitions. End joints in top plates shall be offset at least 24 inches (610 mm). <u>Joints in plates need not occur over studs</u>. Plates shall be a nominal 2 inches in depth (51 mm) and have a width at least equal to the width of the studs.

> **Exception:** A single top plate may be installed in stud walls, provided the plate is adequately tied at joints, corners and intersecting walls by a minimum 3-inch by 6-inch by 0.036-inch-thick (76 mm by 152 mm by 0.914 mm) galvanized steel plate that is nailed to each wall or segment of wall by six 8d nails on each side, provided the rafters or joists are centered over the studs with a tolerance of no more than 1 inch (25.4 mm). The top plate may be omitted over lintels that are adequately tied to adjacent wall sections with steel plates or equivalent as previously described.

CHANGE SIGNIFICANCE. Because of some inconsistencies in the applications of stud wall top-plate joints, the code has been revised to clarify the code and increase consistency. The 2003 IRC section on top

plates requires the joints to be offset but has no criteria for where the joint is placed. Many believe the joints in conventional wood-frame construction to be the weakest link, especially in top plates, which are a critical element of the vertical load-carrying system and of the load-path provisions. On the basis of some previous legacy code figures and other interpretations by various individuals, a segment of the industry has believed that top-plate joints are required to occur on top of a stud. This application would cause the addition of another stud anywhere the joints do not occur over a stud. The 2006 code, by adding a clarifying sentence, makes it clear that top-plate joints are not required to occur over a stud.

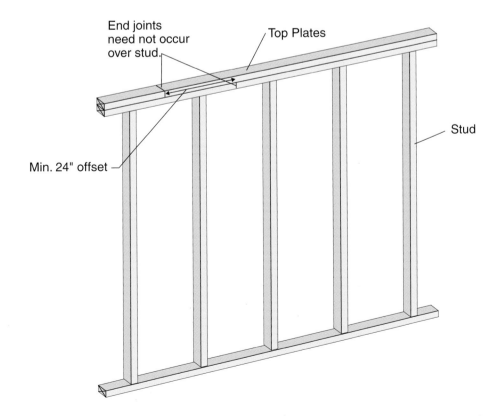

R602.6.1

Drilling and Notching of Top Plate

CHANGE TYPE. Clarification

CHANGE SUMMARY. The code has been revised to clarify that only one metal tie is required to connect double top plates when top plates are cut, notched, or drilled to more than 50% of their width.

2006 CODE: R602.6.1 Drilling and Notching of Top Plate. When piping or ductwork is placed in or partly in an exterior wall or interior load-bearing wall, necessitating cutting, drilling, or notching of the top plate by more than 50 percent of its width, a galvanized metal tie of not less than 0.054 inches thick (1.37 mm) (16 ga) and 1 ½ inches (38 mm) wide shall be fastened ~~to each plate~~ across and <u>to the plate at</u> each side of the opening with not less than eight 16d nails at each side or equivalent. See Figure R602.6.1.

> **Exception:** When the entire side of the wall with the notch or cut is covered by wood structural panel sheathing.

CHANGE SIGNIFICANCE. When top plates are notched, cut, or drilled to more than 50% of their width, thereby severely reducing the cross-sectional area that is needed to carry the applicable loads and jeopardizing the continuity of the structural system, the code requires that galvanized metal ties of at least .054 inch thick and 1.5 inches wide be used to connect the two sides of the top plates weakened by the cut or notch. Cutting, notching, and drilling of top plates are sometimes necessary to accommodate the installation of plumbing pipes or electrical wiring. Many have interpreted this section as a requirement for two such galvanized plates at each cut, where two top plates have been used. This interpretation is partially based on the 2003 code text ". . . shall be fastened to each plate across . . ." and partially based on the fact that when two top plates are present and both are cut, it is reasonable to assume that both plates need to be strengthened. This inter-

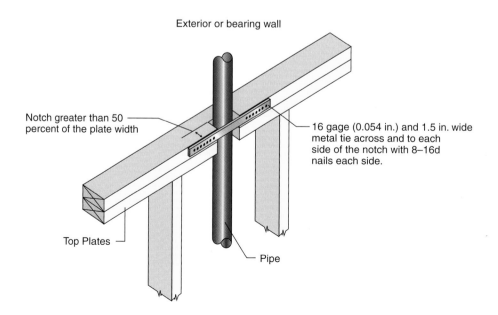

Exterior or bearing wall

Notch greater than 50 percent of the plate width

16 gage (0.054 in.) and 1.5 in. wide metal tie across and to each side of the notch with 8–16d nails each side.

Top Plates

Pipe

pretation, however, appears to be incorrect according to IRC Figure R602.6.1, the IRC Commentary, other interpretations issued by ICC, and the indirect implications of the exception to Section R602.3.2 (the exception allows a single top plate when metal ties are used to tie joints, corners, and intersecting walls). The 2006 code has now removed the phrase "to each plate" and inserted the phrase "the plate" to clarify that even when double top plates are used, only one strap is required to connect the top plates and provide structural continuity. Note that at least eight 16d nails must be used on each side of the metal tie. The picture here shows two metal ties used, each side of each tie using only 4 nails. In such a case, the building official must decide whether the construction is equivalent to the intent of the code.

R602.10.6.1 and Table R602.10.6

Alternate Braced Wall Panels

CHANGE TYPE. Modification

CHANGE SUMMARY. The alternate braced-wall-panel requirements of 2003 code Section R602.10.6 have been assigned to Section R602.10.6.1, and a new table, R602.10.6, has been included to provide options for width and height of braced wall panels.

2006 CODE: R602.10.6.1 Alternate Braced Wall Panels. Alternate braced wall lines constructed in accordance with one of the following provisions shall be permitted to replace each 4 feet (1219 mm) of braced wall panel as required by Section R602.10.4. ~~The maximum height and minimum width of each panel shall be in accordance with Table R602.10.6:~~

1. In one-story buildings, each ~~panel shall have a length of not less than 2 feet, 8 inches (813 mm) and a height of not more than 10 feet (3048 mm). Each~~ panel shall be sheathed on one face with ⅜-inch minimum thickness (10 mm) wood structural

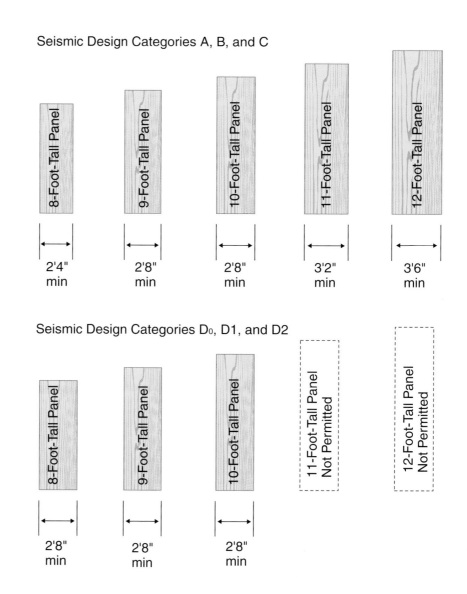

panel sheathing nailed with 8d common or galvanized box nails in accordance with Table R602.3(1) and blocked at all wood structural panel sheathing edges. Two anchor bolts installed in accordance with Figure R403.1(1) shall be provided in each panel. Anchor bolts shall be placed at panel quarter points. Each panel end stud shall have a tie-down device fastened to the foundation, capable of providing an uplift capacity ~~of at least 1800 pounds (816.5 kg)~~ in accordance with Table R602.10.6. The tie-down device shall be installed in accordance with the manufacturer's recommendations. The panels shall be supported directly on a foundation or on floor framing supported directly on a foundation which is continuous across the entire length of the braced wall line. This foundation shall be reinforced with not less than one No. 4 bar top and bottom. When the continuous foundation is required to have a depth greater than 12 inches (305 mm), a minimum 12-inch-by-12-inch (305 mm by 305 mm) continuous footing or turned down slab edge is permitted at door openings in the braced wall line. This continuous footing or turned down slab edge shall be reinforced with not less than one No. 4 bar top and bottom. This reinforcement shall be lapped 15 inches (381 mm) with the reinforcement required in the continuous foundation located directly under the braced wall line.

2. In the first story of two-story buildings, each braced wall panel shall be in accordance with Item 1 above, except that the wood structural panel sheathing shall be provided on both faces, sheathing edge nailing spacing shall not exceed 4 inches (102 mm) on center, at least three anchor bolts shall be placed at one-fifth points~~, and tie-down device uplift capacity shall not be less than 3000 pounds (1360.8 kg)~~.

TABLE R602.10.6   **Minimum Widths and Tie-Down Forces of Alternate Braced Wall Panels**

Seismic Design Category and Windspeed	Tie-Down Force (lb)	Height of Braced Wall Panel				
		Sheathed Width				
		8 ft. 2'–4"	9 ft. 2'–8"	10 ft. 2'–8"	11 ft. 3'–2"	12 ft. 3'–6"
SDC A, B, and C Windspeed ≤ 110 mph	R602.10.6.1, Item 1	1800	1800	1800	2000	2200
	R602.10.6.1, Item 2	3000	3000	3000	3300	3600
		Sheathed Width				
		2'–8"	2'–8"	2'–8"	Note a	Note a
SDC D₀, D₁, and D₂ Windspeed ≤ 110 mph	R602.10.6.1, Item 1	1800	1800	1800	—	—
	R602.10.6.1, Item 2	3000	3000	3000	—	—

For SI: 1 inch = 25.4 mm, 1 foot = 304.8 mm.

a. Not permitted because maximum height is 10 feet.

CHANGE SIGNIFICANCE. The 2003 code contains one alternate braced-wall-panel provision in Section R602.10.6. This alternative method is intended mainly for narrow panels adjacent to wide doors

R602.10.6.1 and Table R602.10.6 continues

R602.10.6.1 and Table R602.10.6
continued

such as garage door openings. The 2006 code contains the same alternate braced wall provisions with some modifications and is assigned a new Section number, R602.10.6.1, because an additional alternate braced-wall methodology has also been added and placed in Section R602.10.6.2.

The changes to the provisions for alternate braced wall panels are intended to provide more flexibility by eliminating the minimum panel length of 2 feet 8 inches and the maximum panel height of 10 feet and inserting a new table, R602.10.6, that allows various panel length and height dimensions. The new table allows narrower widths for shorter panels and allows taller, wider widths in low seismic zones. The width-to-height dimensions found in the table maintain an aspect ratio of $3\frac{1}{2}:1$ in Seismic Design Categories (SDCs) A, B, and C, and no changes have been intended for SDCs D and E. Calculations that were submitted by the proponents of this change show that the shorter, narrower walls will have the same capacity as the 10-foot-tall, 2-foot-8-inchwide wall because wall capacity is governed by the overturning force, which is resisted by a hold-down of defined capacity. As the walls get shorter, there is less overturning force, allowing walls to be narrower, and as the walls get taller, wider panels would be required, as provided in the new table.

CHANGE TYPE.   Addition

CHANGE SUMMARY.   A new section providing alternate bracing methods at door and window openings has been added to the code.

2006 CODE: <u>**R602.10.6.2 Alternate Bracing Wall Panel Adjacent to a Door or Window Opening.**</u>   Alternate braced wall panels constructed in accordance with one of the following provisions are also permitted to replace each 4 feet (1219 mm) of braced wall panel as required by Section R602.10.4 for use adjacent to a window or door opening with a full-length header:

1. In one-story buildings, each panel shall have a length of not less than 16 inches (406 mm) and a height of not more than 10 feet (3048 mm). Each panel shall be sheathed on one face with a single layer of 3/8-inch-minimum-thickness (10 mm) wood structural panel sheathing nailed with 8d common or galvanized box nails in accordance with Figure R602.10.6.2.

R602.10.6.2 continues

R602.10.6.2

Alternate Bracing Wall Panel Adjacent to a Door or Window Opening

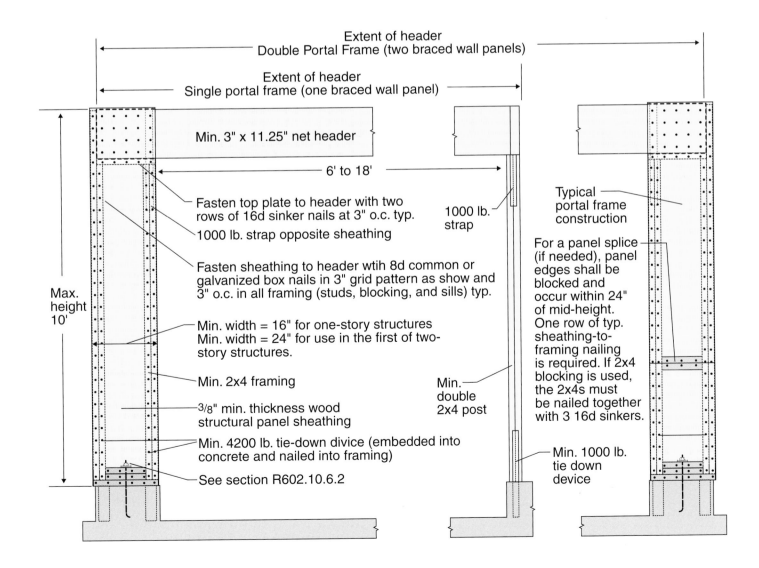

Extent of header
Double Portal Frame (two braced wall panels)

Extent of header
Single portal frame (one braced wall panel)

Min. 3" x 11.25" net header

6' to 18'

Fasten top plate to header with two rows of 16d sinker nails at 3" o.c. typ.

1000 lb. strap opposite sheathing

Fasten sheathing to header wtih 8d common or galvanized box nails in 3" grid pattern as show and 3" o.c. in all framing (studs, blocking, and sills) typ.

Min. width = 16" for one-story structures
Min. width = 24" for use in the first of two-story structures.

Min. 2x4 framing

3/8" min. thickness wood structural panel sheathing

Min. 4200 lb. tie-down divice (embedded into concrete and nailed into framing)

See section R602.10.6.2

Max. height 10'

1000 lb. strap

Min. double 2x4 post

Min. 1000 lb. tie down device

Typical portal frame construction

For a panel splice (if needed), panel edges shall be blocked and occur within 24" of mid-height. One row of typ. sheathing-to-framing nailing is required. If 2x4 blocking is used, the 2x4s must be nailed together with 3 16d sinkers.

R602.10.6.2 continued

The wood structural panel sheathing shall extend up over the solid sawn or glued-laminated header and shall be nailed in accordance with Figure R602.10.6.2. A built-up header consisting of at least two 2 x 12s and fastened in accordance with Table R602.3(1) shall be permitted to be used. A spacer, if used, shall be placed on the side of the built-up beam opposite the wood structural panel sheathing. The header shall extend between the inside faces of the first full-length outer studs of each panel. The clear span of the header between the inner studs of each panel shall be not less than 6 feet (1829 mm) and not more than 18 feet (5486 mm) in length. A strap with an uplift capacity of not less than 1000 pounds (4448 N) shall fasten the header to the side of the inner studs opposite the sheathing. One anchor bolt not less than ⅝-inch-diameter (16 mm) and installed in accordance with Section R403.1.6 shall be provided in the center of each sill plate. The studs at each end of the panel shall have a tie-down device fastened to the foundation with an uplift capacity of not less than 4200 pounds (18,683 N).

Where a panel is located on one side of the opening, the header shall extend between the inside face of the first full-length stud of the panel and the bearing studs at the other end of the opening. A strap with an uplift capacity of not less than 1000 pounds (4448 N) shall fasten the header to the bearing studs. The bearing studs shall also have a tie-down device fastened to the foundation with an uplift capacity of not less than 1000 pounds (4448 N).

The tie-down devices shall be an embedded strap type, installed in accordance with the manufacturer's recommendations. The panels shall be supported directly on a foundation or on floor framing supported directly on a foundation, which is continuous across the entire length of the braced wall line. The foundation shall be reinforced with not less than one No. 4 bar top and bottom.

Where the continuous foundation is required to have a depth greater than 12 inches (305 mm), a minimum 12-inch-by-12-inch (305 mm by 305 mm) continuous footing or turned down slab edge is permitted at door openings in the braced wall line. This continuous footing or turned down slab edge shall be reinforced with not less than one No. 4 bar top and bottom. This reinforcement shall be lapped not less than 15 inches (381 mm) with the reinforcement required in the continuous foundation located directly under the braced wall line.

2. In the first story of two-story buildings, each wall panel shall be braced in accordance with Item 1 above, except that each panel shall have a length of not less than 24 inches (610 mm).

CHANGE SIGNIFICANCE. This is another alternative bracing method allowed to be used adjacent to door and window openings. Four other revisions related to this change and for coordination with it have occurred in Sections R602.10.3, R602.10.4, and R602.10.6.1 and in Table R602.10.5, footnote "c." This new alternative utilizes the header over the adjacent opening where the header runs the length of the sheathed bracing panel—up to the first full-length stud—and is attached to the full-height sheathing with a grid pattern of nails where the sheathing and header overlap. This provides a partially moment-resistant connection at the top connection. At the base of the sheathed section, embedded, nailed framing anchors provide a more conventional movement-resisting connection at the base. These framing anchors are sized to provide uplift in addition to shear capacity, reducing the plate-anchoring requirements. Additional strap anchors, as shown in the figure, are required. The new bracing methods result in reduction of the width of the full-height leg of the assembly from 32 to 16 inches for a one-story building and from 48 inches to 24 inches to the first floor of a two-story building.

This bracing method has been used in certain parts of the country since 1988, and the APA Laboratory has completed a series of cyclic tests based on the Structural Engineers Association of Southern California (SEAOSC) cyclic testing protocol.

Some advantages of this alternative bracing method are the cost-savings associated with eliminating the 48- or 32-inch-wide bracing unit and associated square footage of floor area adjacent to a garage door, and the new panel-width dimensions are more traditional in appearance.

R602.10.11, R602.10.11.1 through R602.10.11.5

Bracing in Seismic Design Categories D₀, D₁, and D₂

CHANGE TYPE.   Modification

CHANGE SUMMARY.   Subsections R602.10.11.1 and R602.10.11.2 of 2003 code Section R602.10.11 have been reformatted into three additional subsections, R602.10.11.1 through R602.10.11.5, for the purpose of improving the clarity of the code. The provisions for exceeding the maximum 25-foot wall bracing and its adjustment factors have been modified by deleting a part of the exception to Section R602.10.11.1 and deleting the 2003 code Table R602.10.11.

2006 CODE: R602.10.11 Bracing in Seismic Design Categories D₀, D₁, and D₂.   Structures located in Seismic Design Categories D₀, D₁, and D₂ shall be provided with exterior and interior braced wall lines.

R602.10.11.1 Braced Wall Line Spacing.   Spacing between braced wall lines in each story shall not exceed 25 feet (7620 mm) on center in both the longitudinal and transverse directions.

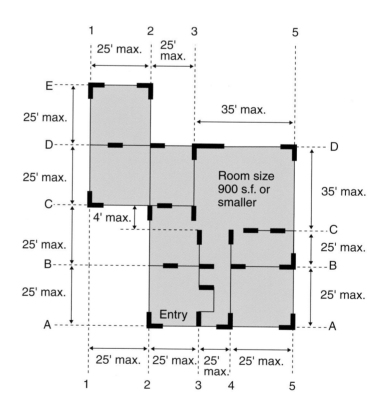

Seismic design categories D_0, D_1, & D_2

-------   Braced wall lines longitudinal A, B, C, D, E

-------   Braced wall lines transverse 1, 2, 3, 4, 5

━━━   Braced wall panels, typical. Amount of panels in each wall based on percentages from Table R602.10.1

Exception: In one- and two-story buildings, spacing between <u>two adjacent</u> braced wall lines shall not exceed 35 feet (10363 mm) on center in order to accommodate one single room not exceeding 900 square feet (84 m2) in each dwelling unit. <u>Spacing between all other braced wall lines shall not exceed 25 feet (7620 mm).</u> ~~The length of wall bracing in braced wall lines spaced greater or less than 25 feet (7620 mm) apart shall be the length required by Table R602.10.1 multiplied by the appropriate adjustment factor from Table R602.10.11.~~

R602.10.11.2 Braced Wall Panel Location.   Exterior braced wall lines shall have a braced wall panel located at each end of the braced wall line.

Exception: For braced wall panel construction Method 3 of Section R602.10.3, the braced wall panel shall be permitted to begin no more than 8 feet (2438 mm) from each end of the braced wall line provided ~~one of~~ the following is satisfied:

1. A minimum 24-inch-wide (610 mm) panel is applied to each side of the building corner and the two 24-inch (610 mm) panels at the corner shall be attached to framing in accordance with Figure R602.10.5; or

2. The end of each braced wall panel closest to the corner shall have a tie-down device fastened to the corner and to the foundation or framing below. The tie-down device shall be capable of providing an uplift allowable design value of at least 1800 pounds (8 KN). The tie-down device shall be installed in accordance with the manufacturer's recommendations.

R602.10.11.3 Collectors.   A designed collector shall be provided if <u>a braced wall panel</u> ~~the bracing~~ is not located at each end of a braced wall line as <u>indicated in Section R602.10.11.2,</u> ~~above~~ or, <u>when using the Section R602.10.11.2 Exception, if a braced wall panel is</u> more than 8 feet (2438 mm) from each end of a braced wall line. ~~as indicated in the exception~~

R602.10.11.<u>4</u> ~~1~~. Cripple Wall Bracing.   In addition to the requirements of Section R602.10.2, where interior braced wall lines occur without a continuous foundation below, the length of parallel exterior cripple wall bracing shall be one and one-half times the length required by Table R602.10.3. Where cripple walls braced using Method 3 of Section R602.10.3 cannot provide this additional length, the capacity of the sheathing shall be increased by reducing the spacing of fasteners along the perimeter of each piece of sheathing to 4 inches (102 mm) on center.

R602.10.11.<u>5</u> ~~2~~. Sheathing Attachment.   Adhesive attachment of wall sheathing shall not be permitted in Seismic Design Categories C, <u>D$_0$,</u> D$_1$, and D$_2$.

R602.10.11, R602.10.11.1 through R602.10.11.5 continues

R602.10.11, R602.10.11.1 through
R602.10.11.5 continued

TABLE R602.10.11 ~~Adjustment of Bracing Amounts for Interior Braced Wall Lines According to Braced Wall Line Spacing~~[a,b]

~~Braced Wall Line Spacing (feet)~~	~~Multiply Bracing Amount in Table R602.10.3 by:~~
~~15 or less~~	~~0.6~~
~~20~~	~~0.8~~
~~25~~	~~1.0~~
~~30~~	~~1.2~~
~~35~~	~~1.4~~

~~For SI 1 foot = 304.8 mm~~

~~a. Linear interpretation is permissible.~~
~~b. The adjustment is limited to the larger spacing between braced wall lines to either side of an interior braced wall.~~

CHANGE SIGNIFICANCE. A major part of the revisions resulting in the 2006 code in this section is reformatting and editorial changes that were intended to make the application of the code easier and clearer without any technical implications. As a result, the 2003 code Section R602.10.11 and its subsections R602.10.11.1 and R602.10.11.2, as well as the various exceptions, have all been reorganized into the new Section R602.10.11 and five subsections, R602.10.11.1 through R602.10.11.5.

The technical modification of this section is related to the spacing provisions of braced wall lines. The maximum spacing of braced wall lines is required to be 25 feet, with an exception in one- and two-story buildings to accommodate one single room of maximum 900 square feet. In the 2003 code, this exception includes provisions for braced wall line spacings that could be less than or greater than 25 feet, provided an adjustment factor found in Table R602.10.11 is applied. The 2006 code no longer contains this part of the exception and consequently has deleted Table R602.10.11 as well.

The text and table deletions discussed above result in elimination of a reduction in the amount of bracing allowed for interior braced wall lines spaced less than 25 feet, which is a part of 2003 code Table R602.10.11. Additionally, this will allow the spacing of interior as well as exterior braced wall lines (2003 code Table R602.10.11 applied to interior braced wall lines only) to exceed 25 feet and go up to 35 feet in one- and two-story buildings for a single room of 900 square feet without having to increase the amount of bracing that resulted from the application of the adjustment factor. These changes have made the application of braced wall line spacing provisions simpler and have expanded it to include the exterior walls when the exception for large rooms is used.

R602.11.1
Wall Anchorage

CHANGE TYPE. Addition

CHANGE SUMMARY. An alternative method for diagonally slotted holes is provided to resist seismic forces.

2006 CODE: R602.11.1 Wall Anchorage. Braced wall line sills shall be anchored to concrete or masonry foundations in accordance with Sections R403.1.6 and R602.11. For all buildings in Seismic Design Categories D_0, D_1, and D_2 and townhouses in Seismic Design Category C, plate washers, a minimum of ~~¼~~ 0.229 inch by 3 inches by 3 inches (5.8 mm by 76 mm by 76 mm) in size, shall be installed between the foundation sill plate and the nut. The hole in the plate washer is permitted to be diagonally slotted with a width of up to ³⁄₁₆ inch (5 mm) larger than the bolt diameter and a slot length not to exceed 1¾ inches (44 mm), provided a standard cut washer is placed between the plate washer and the nut.

CHANGE SIGNIFICANCE. The plate washer thickness has been changed from ¼" nominal (0.25 inch) to a 0.229 inch actual thickness. Even though this is a reduction in washer thickness, it is believed to still be able to carry the required loads and is much more readily available. Additionally, because of some construction tolerance issues for 3 by 3 by 0.229-inch plate washers, an alternative method of diagonal holes has now been provided in the code. This new text permits use of diagonal slotted holes in the plate washers to address such construction tolerance issues. The language for the slotted holes is taken directly from 2003 National Earthquake Hazards Reduction Program (NEHRP) provisions. The Wood Frame Project found that in order to reduce sill splitting, it is critical to get the edge of the plate washer very close to the edge of the sill at which braced wall sheathing is fastened. Tests with larger distances between edge of sill and edge of

R602.11.1 continues

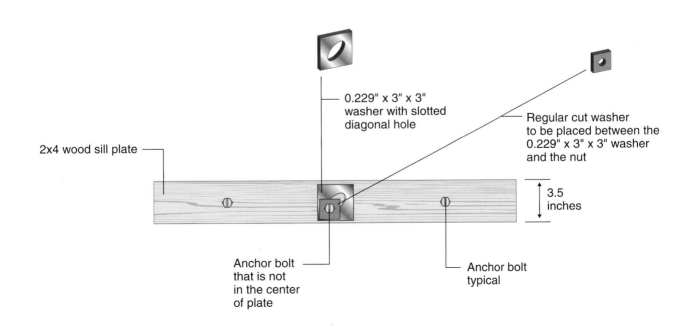

0.229" x 3" x 3" washer with slotted diagonal hole

Regular cut washer to be placed between the 0.229" x 3" x 3" washer and the nut

2x4 wood sill plate

3.5 inches

Anchor bolt that is not in the center of plate

Anchor bolt typical

R602.11.1 continued   washer revealed splitting at significantly lower force levels. The diagonal slotted hole provisions in the 2006 IRC were provided in lieu of the original proposal that attempted to reduce the washer sizes to 2-½ by 2-½ by $\frac{3}{16}$ inches. Because washers of minimum size 3 by 3 by 0.229 inch are needed, the diagonal slotted holes were approved for inclusion in the code.

R603.3.2
Load-Bearing Walls

CHANGE TYPE. Addition

CHANGE SUMMARY. The loading application of steel cold-formed walls has been clarified, and new prescriptive tables have been added to show cold-formed-steel stud span and thickness for bearing walls.

2006 CODE: R603.3.2 Load-Bearing Walls. Steel studs shall comply with Tables R603.3.2(2) through ~~R603.3.2(7)~~ R603.3.2(21). ~~for steels with minimum yield strength of 33 ksi (227.7 MPa) and Tables R603.3.2(8) through R603.3.2(13) for steels with minimum yield strength of 50 ksi (345 MPa)~~. The tabulated stud thickness for structural walls shall be used when the attic load is 10 psf (0.48 KPa) or less. When an attic storage load is greater than 10 psf (0.48 KPa) but less than or equal to 20 psf (0.96 KPa) the design value the next higher snow load column value from Tables R603.3.2(2) through R603.3.2(21) shall be used to select the stud size. The tabulated stud thickness for structural walls supporting one floor, roof, and ceiling shall be used when the second-floor live load is 30 psf (1.44 KPa). When the second-floor live load is greater than 30 psf (1.44 KPa) but less than or equal to 40 psf (1.92 KPa) the design value in the next higher snow load column from Tables R603.2(12) through R603.3.2(21) shall be used to select the stud size.

Fastening requirements shall be in accordance with Section R603.2.4 and Table R603.3.2(1). Tracks shall have the same minimum thickness as the wall studs. Exterior walls with a minimum of ½-inch (13 mm) gypsum board installed in accordance with Section R702 on the interior surface and wood structural panels of minimum $^{7}/_{16}$-inch thick (11 mm) oriented-strand board or $^{15}/_{32}$-inch thick (12 mm) plywood installed in accordance with Table R603.3.2(1) on the outside surface shall be permitted to use the next thinner stud from Tables R603.3.2(2) through R603.3.2(13) but not less than 33 mils (0.84 mm). Interior load-bearing walls with a minimum ½-inch (13 mm) gypsum board installed in accordance with Section R702 on both sides of the wall shall be permitted to use the next thinner stud from Tables R603.3.2(2) through R603.3.2(13) but not less than 33 mils (0.84 mm).

CHANGE SIGNIFICANCE. Section R603.3.2 and its related tables provide prescriptive stud-span and thickness information for exterior and load-bearing walls. The 13 span tables of the 2003 IRC code have been completely reformatted and reorganized into 21 new tables that provide the needed information in a more user-friendly and easy-to-follow format. The new tables are organized according to building width and elements being supported (for example, Table R603.3.2(21) is for 40-foot-wide buildings supporting one floor, roof, and ceiling). Additionally, specific criteria have been added to enable using the table to determine thickness values when attics have storage loads of less than or equal to 10 psf or more than 10 psf. Similar criteria have been added for floors with different live loads.

R603.3.2 continues

R603.3.2 continued

TABLE R603.3.2(21)   40-Foot Wide Building Supporting One Floor, Roof, and Ceiling[a,b,c] 50 ksi Steel

Wind Speed Exp A/B	Exp C	Member Size	Stud/Spacing (inch)	8-Foot Studs				9-Foot Studs				10-Foot Studs			
				20	30	50	70	20	30	50	70	20	30	50	70
85 mph		350S162	16	33	33	33	33	33	33	33	33	33	33	33	33
			24	43	43	43	54	43	43	43	43	43	43	43	54
		550S162	16	33	33	33	33	33	33	33	33	33	33	33	33
			24	33	33	43	43	33	33	33	33	33	33	33	43
90 mph		350S162	16	33	33	33	33	33	33	33	33	33	33	33	33
			24	43	43	43	54	43	43	43	43	43	43	43	54
		550S162	16	33	33	33	33	33	33	33	33	33	33	33	33
			24	33	33	43	43	33	33	33	33	33	33	33	43
100 mph	85 mph	350S162	16	33	33	33	33	33	33	33	33	33	33	33	33
			24	43	43	43	54	43	43	43	43	43	43	43	54
		550S162	16	33	33	33	33	33	33	33	33	33	33	33	33
			24	33	33	43	43	33	33	33	33	33	33	33	43
110 mph	90 mph	350S162	16	33	33	33	33	33	33	33	33	33	33	33	43
			24	43	43	43	54	43	43	43	43	43	43	43	54
		550S162	16	33	33	33	33	33	33	33	33	33	33	33	33
			24	33	33	43	43	33	33	33	33	33	33	33	43
	100 mph	350S162	16	33	33	33	33	33	33	33	33	33	33	33	43
			24	43	43	43	54	43	43	43	43	43	54	54	54
		550S162	16	33	33	33	33	33	33	33	33	33	33	33	33
			24	33	33	43	43	33	33	33	33	33	33	33	43
	110 mph	350S162	16	33	33	33	43	33	33	33	43	33	43	43	43
			24	43	43	43	54	43	43	43	54	54	54	54	54
		550S162	16	33	33	33	33	33	33	33	33	33	33	33	33
			24	33	33	43	43	33	33	43	43	33	33	43	43

For SI: 1 inch = 25.4 mm, 1 foot = 304.8 mm, 1 mil = 0.0254 mm, 1 mile per hour = 0.447 m/s, 1 pound per square foot = 0.0479 KPa, 1 ksi = 1000 psi = 6.895 MPa.

a. Deflection criteria: L/240
b. Design load assumptions:
    Second floor dead load is 10 psf
    Second floor live load is 30 psf
    Roof/ceiling dead load is 12 psf
    Attic live load is 10 psf
c. Building width is in the direction of horizontal framing members supported by the wall studs.

R606.3 and R606.4

Corbelled Masonry

CHANGE TYPE. Addition

CHANGE SUMMARY. An alternative method of using ties in lieu of headers in corbelled masonry has been provided in Section 606.3, and another new section, 606.4.1, has been added that provides the requirements for masonry support.

2006 CODE: R606.3 Corbelled Masonry. Solid masonry units shall be used for corbelling. The maximum corbelled projection beyond the face of the wall shall not be more than one-half of the wall thickness or one-half the wythe thickness for hollow walls; the maximum projection of one unit shall not exceed one-half the height of the unit or one-third the thickness at right angles to the wall. ~~The top course of corbels shall be a header course w~~ When ~~the~~ corbelled masonry is used to support floor or roof-framing members <u>the top course of the corbel shall be a header course or the top course bed joint shall have a ties to the vertical wall.</u> ~~projection~~ <u>The hollow space behind the corbelled masonry shall be filled with mortar or grout.</u>

~~R606.3.1~~ <u>R606.4</u> Support Conditions. <u>Bearing and support conditions shall be in accordance with Sections R606.4.1 and R606.4.2.</u>

R606.4.1 Bearing on Support. <u>Each masonry wythe shall be supported by at least ⅔ of the wythe thickness.</u>

R606.4.2 Support at Foundation. Cavity wall or masonry veneer construction may be supported on an 8-inch (203 mm) foundation

R606.3 and R606.4 continues

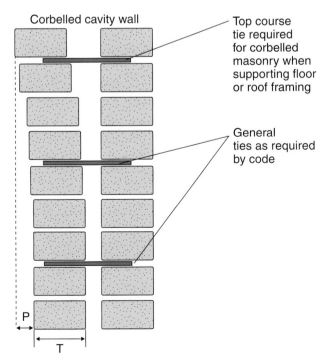

Corbelled cavity wall

Top course tie required for corbelled masonry when supporting floor or roof framing

General ties as required by code

P

T

Projection P ≤ ¹/₂ T

R606.3 and R606.4 continued

wall, provided the 8-inch (203 mm) wall is corbelled with solid masonry to the width of the wall system above. The total horizontal projection of the corbel shall not exceed 2 inches (51 mm) with individual corbels projecting not more than one-third the thickness of the unit or one-half the height of the unit. ~~The top course of all corbels shall be a header course~~.

CHANGE SIGNIFICANCE. The top course of corbelled masonry construction must be able to transfer the supporting loads into the wall section. This requirement is applicable when the corbelled masonry is used to support floor or roof-framing members. The load transfer is accomplished by requiring that the top course be a header. In the 2006 code another method of transferring such loads has been included, so that either the header procedure or using ties in the bed joint of the top course to transfer the loads to the vertical portion of the wall is acceptable. The tie characteristics or spacing has not been provided in this section; therefore, the requirements from elsewhere in the code should be used. Subsection R606.3.1 has now been relocated to new Section R606.4 and titled "support conditions" so that it is now a general requirement that applies to all masonry construction rather than being hidden under the corbelled masonry section. The section reorganization was necessary because no general masonry support requirements were found in the 2003 code, and it was not appropriate to insert such criteria in the corbelled masonry section. The new Section R606.4.2 now requires that each masonry wythe have no more than $\frac{1}{3}$ of its width extending beyond the solid support.

CHANGE TYPE.   Clarification and Addition

CHANGE SUMMARY.   The table headings have been modified to clarify the application of nominal versus actual wall sizes, and a new footnote has been added to allow 3.5-inch flat ICF walls to be constructed 1 inch thicker.

2006 CODE:

Table R611.2
Requirements for ICF Walls

TABLE R611.2   Requirements for ICF Walls[a,b,c]

Wall Type and ~~Nominal~~ Size	Maximum Wall Weight (psf)[c]	Minimum Width of Vertical Core (inches)[a]	Minimum Thickness of Vertical Core (inches)[a]	Maximum Spacing of Vertical Cores (inches)	Maximum Spacing of Horizontal Cores (inches)	Minimum Web Thickness (inches)
3.5″ Flat	44[d]	N/A	N/A	N/A	N/A	N/A
5.5″ Flat	69	N/A	N/A	N/A	N/A	N/A
7.5″ Flat	94	N/A	N/A	N/A	N/A	N/A
9.5″ Flat	119	N/A	N/A	N/A	N/A	N/A
6″ Waffle-Grid	56	6.25	5	12	16	2
8″ Waffle-Grid	76	7	7	12	16	2
6″ Screen-Grid	53	5.5	5.5	12	12	N/A

For SI: 1 inch = 25.4 mm; 1 pcf = 16.018 kg/m^3; 1 psf = 0.0479 KPa.

a. For width "W", thickness "T", spacing, and web thickness, refer to Figures R611.4 and R611.5.
b. N/A indicates not applicable.
c. Wall weight is based on a unit weight of concrete of 150 pcf (23.6 kN/m3). The tabulated values do not include any allowance for interior and exterior finishes.
d. For all buildings in Seismic Design Category A or B, and detached one- and two-family dwellings in Seismic Design Category C, the actual wall thickness is permitted to be up to one-inch (25.4 mm) thicker than shown and the maximum wall weight to be 56 psf. Construction requirements and other limitations within Section R611 for 3.5-inch flat ICF walls shall apply. Interpolation between provisions for 3.5-inch and 5.5-inch flat ICF walls is not permitted.

CHANGE SIGNIFICANCE.   The table has been confusing in relation to the application of the wall size. Whereas the flat wall provisions in Section 611 are based on the actual dimensions of wall assemblies and the waffle-grid and screen-grid provisions are based on nominal sizes, the heading of the first column assigned nominal sizes to all of these systems. The 2006 code heading for the first column has deleted the word "Nominal" to clarify this issue. Further clarifications in the table include the relocation of Footnote references "a" and "c" from the table title because they were not applicable to the entire table.

The addition of the new footnote "d" creates provisions for ICF walls that do not exactly fit within the table values, because some ICF manufacturers produce forms that result in walls 4.5 inches thick. The footnote allows for such 4.5-inch-thick walls to be constructed to the requirements of 3.5-inch walls in the lower SDCs of A, B, and C. The use of 3.5-inch provisions for 4.5-inch-thick walls that are heavier, although advantageous for wind design and uplift resistance, is less advantageous for seismic design because of the increase in seismic weight that

Table R611.2 continues

Table R611.2 continued

results in higher base shear values. To deal with this issue, the footnote requires the maximum wall weight of 56 psf, which is the interpolation between the 3.5- and 5.5-inch dimensions (44 + 69 ÷ 2 = 56.5), but it does not allow any other interpolations in various tables and subsections of Section 611.

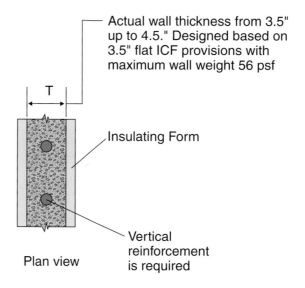

Actual wall thickness from 3.5"
up to 4.5." Designed based on
3.5" flat ICF provisions with
maximum wall weight 56 psf

T

Insulating Form

Vertical
reinforcement
is required

Plan view

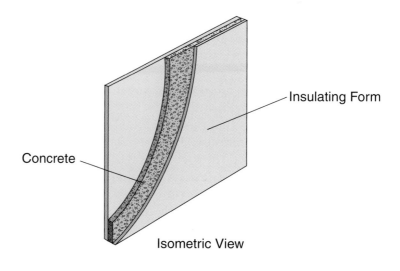

Concrete

Insulating Form

Isometric View

CHANGE TYPE. Addition

R613.1
General Window Installation Instructions

CHANGE SUMMARY. The new text requires window manufacturers to provide installation instructions and requires the installation to comply with the product's installation instructions.

2006 CODE: R613.1 General. This section prescribes performance and construction requirements for exterior window systems installed in wall systems. ~~Waterproofing, sealing and flashing systems are not included in the scope of this section.~~ <u>Windows shall be installed and flashed in accordance with the manufacturers written installation instructions. Written installation instructions shall be provided by the manufacturer for each window.</u>

CHANGE SIGNIFICANCE. Section R613, titled exterior windows and glass doors, contains mostly structural-loading, anchoring, and testing requirements. The 2003 code excludes any provisions related to waterproofing, sealing, and flashing systems by specifically saying so in the "General" section. In addition to their structural performance, exterior windows and glass doors are important elements for protecting the building interior from exterior elements such as water protrusion and adverse temperatures. The lack of any provisions for proper installation of exterior windows and glass doors leaves the building official without the tools to seek corrective action, even in cases where such elements have clearly been installed improperly. The 2006 code has removed this exclusion and has added language that each exterior window must be provided with the manufacturer's installation instructions and be installed and flashed according to such instructions.

R613.2
Window Sills

CHANGE TYPE.   Addition

CHANGE SUMMARY.   Provisions have been added for window sill height and protection of fall from exterior operable windows when such windows are more than 72 inches above exterior finished grade.

2006 CODE: <u>**R613.2 Window Sills.** In dwelling units, where the opening of an operable window is located more than 72 inches (1829 mm) above the finished grade or surface below, the lowest part of the clear opening of the window shall be a minimum of 24 inches (610 mm) above the finished floor of the room in which the window is located. Glazing between the floor and 24 inches (610 mm) shall be fixed or have openings such that a 4 inch (102 mm) diameter sphere cannot pass through.</u>

<u>**Exceptions:**</u>
<u>1. Windows whose openings will not allow a 4 inch diameter (102 mm) sphere to pass through the opening when the opening is in its largest opened position.</u>

<u>2. Openings that are provided with window guards that comply with ASTM F2006 or F2090.</u>

CHANGE SIGNIFICANCE.   This new section is intended to reduce the number of falls through exterior windows for small children. Windows adjacent to exterior finished grade with more than a 72-inch difference

in elevation from the window opening to such finished grade must have sills to at least 24 inches above the floor of the room where the window is located or the window opening must be protected with devices similar to a guardrail that will prevent the passage of a 4-inch sphere. The statistics provided in support of this code change estimated that about 1000 children fall each year from exterior windows because of low windowsills. This issue has been hotly debated for many years in the code development arena; proponents believe that increasing the sill height will prevent many serious injuries and deaths among small children and will save an enormous amount in medical bills each year, whereas opponents believe that there is no substantiation that low windowsills are the cause of children falling and that emergency egress will be adversely affected by higher windowsills. The new language is applicable to every window in exterior walls, regardless of the room in the dwelling where the window is located, and as such it will need additional attention with regard to the emergency egress and rescue windows required in sleeping rooms. The maximum windowsill height for emergency escape and rescue openings is 44 inches above the floor, per Section R310.1, and there is no minimum sill height requirement. As a result, if the option of raising the windowsills to 24 inches is used for fall prevention, the emergency egress windows will not be affected in any way. However, if the option of providing devices that prevent passage of a 4-inch sphere is used, as allowed by the exception, then such devices must be releasable or removable from the inside without the use of tools, special knowledge, or effort, as specified in Section R310.4. The new ASTM standards F2006 and F2090, referenced in the exception, are standards for guards for non-emergency escape and emergency escape windows respectively.

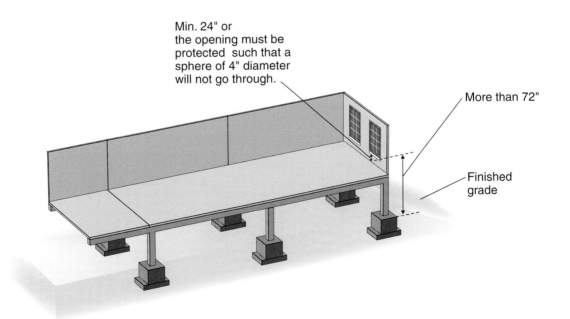

Min. 24" or the opening must be protected such that a sphere of 4" diameter will not go through.

More than 72"

Finished grade

R613.7

Wind-borne Debris Protection

CHANGE TYPE.   Clarification

CHANGE SUMMARY.   The protection of exterior windows and exterior glass doors from windborne debris has been revised to be required in windborne debris regions rather than hurricane-prone regions.

2006 CODE: R613.7 Wind-borne Debris Protection.   Protection of exterior windows and glass doors in buildings located in ~~hurricane-prone~~ <u>wind-borne debris</u> regions from windborne debris shall be in accordance with Section R301.2.1.2.

CHANGE SIGNIFICANCE.   Both of the terms *hurricane-prone regions* and *windborne debris regions* are defined in Section R202. Whereas hurricane-prone regions consist of large geographical areas vulnerable to hurricanes where the basic wind speed is greater than 90 mph, windborne debris regions are limited areas within hurricane-prone regions within 1 mile of the coastal high-water line where the basic wind speed is 110 mph or greater. Windborne debris damage is a problem in all hurricane-prone regions, but it is most severe with higher wind speeds and close to coastal lines. The 2003 code references windborne debris protection in all hurricane-prone regions, in accordance with Section R301.2.1.2, which in turn requires windborne de-

	Location	Vmph
90	Hawaii	105
100	Puerto Rico	145
110	Guam	170
120	Virgin Islands	145
130	American Samoa	125
140		
150		
Special Wind Region		

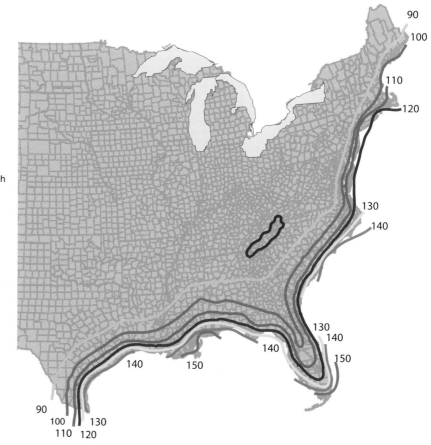

bris protection only in windborne debris regions. The apparent conflict between these two sections caused some to wonder whether windborne debris protection is required in all hurricane-prone regions. It had never been intended that this requirement be applied to the vast geographical areas that are merely prone to hurricanes. The revised language clarifies that windborne debris protection is required only in windborne debris regions.

R613.7.1

Fenestration Testing and Labeling

CHANGE TYPE. Addition

CHANGE SUMMARY. The new section requires testing and labeling for windows in windborne debris regions and provides provisions for third-party testing.

2006 CODE: <u>**R613.7.1 Fenestration Testing and Labeling.** Fenestration shall be tested by an approved independent laboratory, listed by an approved entity, and bear a label identifying manufacturer, performance characteristics, and approved inspection agency to indicate compliance with the requirements of the following specifications:</u>

<u>**1.** ASTM E1886 and ASTM E1996 or</u>
<u>**2.** AAMA 506</u>

CHANGE SIGNIFICANCE. Section R613.3 contains provisions for testing and labeling of exterior windows and glass doors. This is a generic requirement, and as such it is used to require testing for all such exterior windows and glass doors in all regions, including windborne debris regions. A review of standards referenced within Section R613.3, however, reveals that these standards are not applicable and appropriate for specific requirements of the windborne debris regions. The new Section R613.4.1, a subsection related to windborne debris protection, introduces new language for testing and labeling of fenestration in windborne debris regions and makes reference to appropriate standards of ASTM E 1886 and ASTM E1996 or American Architectural Manufacturer's Association (AAMA) 506.

This revision makes this section for windborne debris consistent with the text in R613.3 that specifically identifies the standard for window compliance in non-windborne debris openings. It also acknowledges a third-party certification program that became available for certification in 2002. This program requires testing in accredited independent laboratories, review by an accredited independent body, and random in-plant inspections. The program includes a permanent label for easy field identification of the product. The AAMA 506-02 document requires successful ASTM E1886 and ASTM E1996 test results.

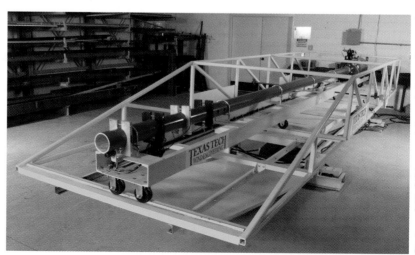

R613.9.1
Mullions

CHANGE TYPE. Modification

CHANGE SUMMARY. The modified language requires that testing of mullions be done in accordance with AAMA 450 and makes the criteria for actual tests of entire mullion assemblies consistent with IBC Chapter 24.

2006 CODE: R613.9.1 Mullions. Mullions shall be tested by an approved testing laboratory in accordance with AAMA 450, or be engineered in accordance with accepted engineering practice. ~~Both methods~~ Mullions tested as stand alone units or qualified by engineering shall use performance criteria cited in Sections R613.9.2, R613.9.3 and R613.9.4. Mullions qualified by an actual test of an entire assembly shall comply with R613.9.2 and R613.9.4.

CHANGE SIGNIFICANCE. This modification requires that the testing of mullions used as support between glazed assemblies in residential construction be done in accordance with AAMA 450. AAMA 450 pulls together all the current requirements for mullions, puts them all in one place, and gives clear pass/fail and rating criteria. Additionally, new language for testing of entire mullion assemblies has been introduced that eliminates the deflection criteria of Section R613.6.3 for such assemblies. The deflection limit of Section R613.6.3 is a neces-

R613.9.1 continues

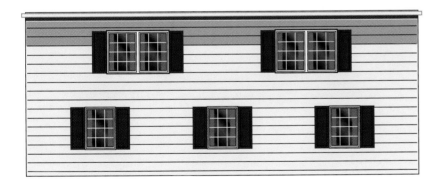

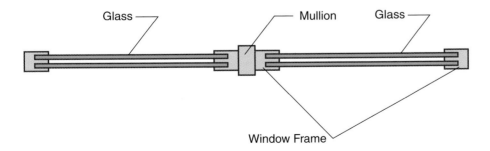

Glass — Mullion Glass —

Window Frame

Top view

R613.9.1 continued   sary requirement to establish a limiting condition for an engineering calculation or where the mullion has been individually tested. Such a calculation needs this boundary to compensate for influences of dimensional and material variability, tolerance stack-up, and other unknown field conditions. The tested assembly accomplishes this with an actual test specimen that is subjected to 150% of its design load; thus, deflection criteria are no longer necessary.

CHANGE TYPE.   Addition

CHANGE SUMMARY.   New provisions for use of ceiling gypsum board construction in resisting lateral forces have been introduced. This addition makes the IRC consistent with the IBC, which contains the same provisions in slightly different format in Section 2508.5.

2006 CODE: **R702.3.7 Horizontal Gypsum Board Diaphragm Ceilings.**   Use of gypsum board shall be permitted on wood joists to create a horizontal diaphragm in accordance with Table R702.3.7. Gypsum board shall be installed perpendicular to ceiling framing members. End joints of adjacent courses of board shall not occur on the same joist. The maximum allowable diaphragm proportions shall be $1\frac{1}{2}$:1 between shear resisting elements. Rotation or cantilever conditions shall not be permitted. Gypsum board shall not be used in diaphragm ceilings to resist lateral forces imposed by masonry or concrete construction. All perimeter edges shall be blocked using wood members not less than 2-inch (51 mm) by 6-inch (152 mm) nominal dimension. Blocking material shall be installed flat over the top plate of the wall to provide a nailing surface not less than 2 inches (51 mm) in width for the attachment of the gypsum board.

R702.3.7

Horizontal Gypsum Board Diaphragm Ceilings

TABLE R702.3.7

Shear Capacity for Horizontal Wood-Framed Gypsum Board Diaphragm Ceiling Assemblies

Material	Thickness of Material (min.) (in.)	Spacing of Framing Members (max.) (in.)	Shear Value[a,b] (plf of ceiling)	Minimum Fastener Size[c,d]
Gypsum Board	$\frac{1}{2}$	16 o.c.	90	5d cooler or wallboard nail; $1\frac{5}{8}$-inch long; 0.086 inch shank; $\frac{15}{64}$ inch head
Gypsum Board	$\frac{1}{2}$	24 o.c.	70	5d cooler or wallboard nail; $1\frac{5}{8}$-inch long; 0.086 inch shank; $\frac{15}{64}$ inch head

For SI: 1 inch = 25.4 mm, 1 pound per linear foot = 1.488 kg/m

a. Values are not cumulative with other horizontal diaphragm values and are for short-term loading due to wind or seismic loading. Values shall be reduced 25% for normal loading.
b. Values shall be reduced 50% in Seismic Categories D_0, D_1, D_2, and E.
c. $1\frac{1}{4}$"; #6 Type S or W screws may be substituted for the listed nails.
d. Fasteners shall be spaced not more than 7 inches on center at all supports, including perimeter blocking, and not less than $\frac{3}{8}$ inch from the edges and ends of the gypsum board.

CHANGE SIGNIFICANCE.   New provisions mirroring those of the IBC have been included in the IRC to allow gypsum board to be used in diaphragms to resist lateral shear forces. Gypsum board is allowed to be used as a membrane in horizontal diaphragm ceilings. When installed according to specific parameters, it provides another method of creating a finished surface that can also resist horizontal shear and wind forces. The section prohibits the use of gypsum board in diaphragm ceilings to resist lateral forces imposed by masonry or concrete. This is because gypsum board systems lack sufficient rigidity and there are significant differences in in-plane stiffness of masonry or concrete systems and horizontal wood diaphragms incorporating ■ gypsum board.

R702.3.7 continues

R702.3.7 continued

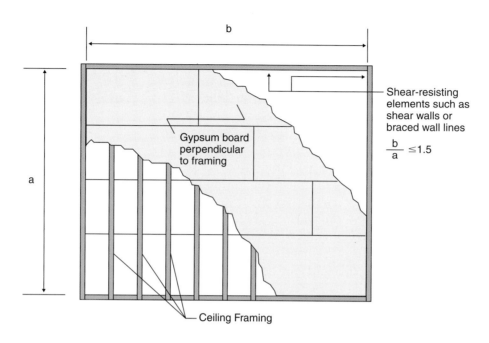

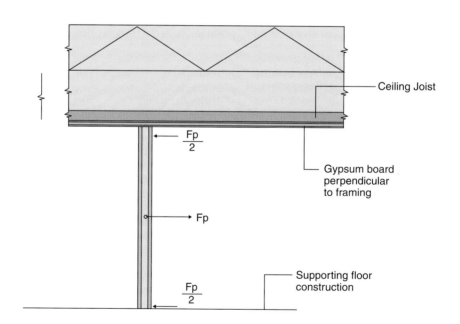

CHANGE TYPE.   Addition

CHANGE SUMMARY.   "Green gypsum board" is no longer allowed to be used as backer behind tiled tub and shower walls.

2006 CODE: <u>**R702.4.2. Cement, Fiber-Cement, and Glass Mat Gypsum Backers.**</u> <u>Cement, fiber-cement, or glass mat gypsum backers in compliance with ASTM C 1288, C 1325, or C1178 and installed in accordance with manufacturer recommendations shall be used as backers for wall tile in tub and shower areas and wall panels in shower areas.</u>

CHANGE SIGNIFICANCE.   The 2003 IRC does not prohibit the use of gypsum "green board" in areas behind shower and tub walls. The new section R702.4.2 prohibits the use of green board by requiring three specific materials for backing in tiled tub and shower walls: cement board, fiber-cement board, and glass mat gypsum backers. In tiled tub and shower walls, because of the possibility of cracks in the grout joints or deteriorated caulking, the water could penetrate behind the tiles and be absorbed by the paper facing of the green board.

R702.4.2 continues

R702.4.2

Cement, Fiber-Cement, and Glass Mat Gypsum Backers

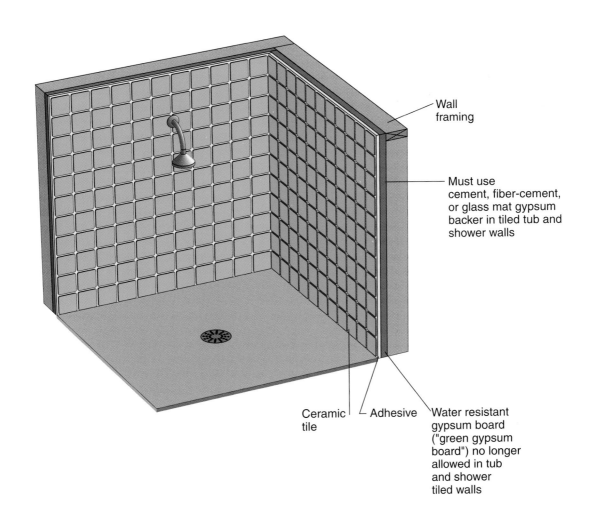

Wall framing

Must use cement, fiber-cement, or glass mat gypsum backer in tiled tub and shower walls

Ceramic tile

Adhesive

Water resistant gypsum board ("green gypsum board") no longer allowed in tub and shower tiled walls

R702.4.2 continued   Application of water for prolonged periods to the surface of the green board can cause the paper facing to delaminate and over time cause the failure of the system as intended for protection of building elements. Application of gypsum board backers in other areas and ceilings is regulated in Section R702.3.8.

CHANGE TYPE.   Modification

CHANGE SUMMARY.   This modification clarifies that a means of draining water to the exterior must be provided for exterior wall assemblies and includes provisions for alternative methods of exterior-wall-assembly water management through the exceptions. This change makes the IRC more consistent with the IBC.

2006 CODE: R703.1 General.   Exterior walls shall provide the building with a weather-resistant exterior wall envelope. The exterior wall envelope shall include flashing as described in Section R703.8. The exterior wall envelope shall be designed and constructed in such a manner as to prevent the accumulation of water within the wall assembly by providing a water-resistive barrier behind the exterior veneer as required by Section R703.2 <u>and a means of draining water that enters the assembly to the exterior. Protection against condensation in the exterior wall assembly shall be provided in accordance with Chapter 11 of this code.</u>

R703.1 continues

R703.1

General Draining Exterior Wall Assemblies

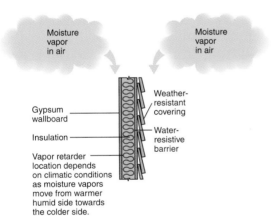

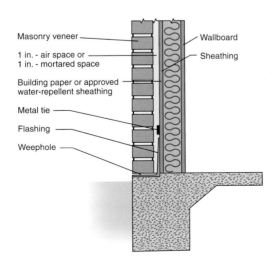

R703.1 continued

Exceptions:

1. A weather-resistant exterior wall envelope shall not be required over concrete or masonry walls designed in accordance with Chapter 6 and flashed according to Section R703.7 or R703.8.

2. Compliance with the requirements for a means of drainage, and the requirements of Section 703.2 and Section 703.8, shall not be required for an exterior wall envelope that has been demonstrated to resist wind-driven rain through testing of the exterior wall envelope, including joints, penetrations and intersections with dissimilar materials, in accordance with ASTM E331 under the following conditions:

   2.1 Exterior wall envelope test assemblies shall include at least one opening, one control joint, one wall/eave interface and one wall sill. All tested openings and penetrations shall be representative of the intended end-use configuration.

   2.2 Exterior wall envelope test assemblies shall be at least 4 feet (1219 mm) by 8 feet (2438 mm) in size.

   2.3 Exterior wall assemblies shall be tested at a minimum differential pressure of 6.24 pounds per square foot (299 Pa).

   2.4 Exterior wall envelope assemblies shall be subjected to a minimum test exposure duration of 2 hours.

The exterior wall envelope design shall be considered to resist wind-driven rain where the results of testing indicate that water did not penetrate control joints in the exterior wall envelope, joints at the perimeter of openings penetration, or intersections of terminations with dissimilar materials.

CHANGE SIGNIFICANCE. The building exterior-wall envelopes are susceptible to water penetration and the possibility of damage to the wall assembly and interior building elements if the penetrated water and associated moisture issues are not dealt with effectively. Moisture either should be prevented from entering the wall assembly or must be drained; otherwise, moisture build-up can lead to rapid deterioration of the wall assembly and possibly the structural members. Various methods of flashing, providing space between veneer and wall face, weep holes, and other methods of dealing with water and moisture have been scattered in different sections of the code. The modifications in this section introduce new language to reinforce what should already be applied through other sections of the code (for example, flashing in Section R703.8) and provide alternative means and methods of dealing with moisture issues in exterior walls. The new performance language requires that any water entering the exterior wall assembly must be drained to the exterior. The modification also makes the IRC requirements more consistent with the same provisions in the IBC, which are covered in Section 1403.2.

CHANGE TYPE.   Modification

CHANGE SUMMARY.   The felt paper required to be applied over exterior wall studs or sheathing must be continuous to the top of walls, and there is no longer an exception for not requiring such felt paper under panel sidings with shiplap joints. Table R703.4 has been modified to require a weather-resistive barrier under all exterior siding materials.

2006 CODE: R703.2 ~~Weather-Resistant Sheathing Paper.~~ Water-Resistive Barrier.   ~~Asphalt saturated felt free from holes and breaks, weighing not less tha14 pounds per 100 square feet (0.683 kg/m^2) and complying with ASTM D 226 or other approved weather-resistant material shall be applied over studs or sheathing of all exterior walls as required by Table R703.4.~~ One layer of No. 15 asphalt felt, free from holes and breaks, complying with ASTM D 226 for Type 1 felt or other approved water-resistive barrier shall be applied over studs or sheathing of all exterior walls. Such felt or material shall be applied horizontally, with the upper layer lapped over the lower layer not less than 2 inches (51 mm). Where joints occur, felt shall be lapped not less than 6 inches (152 mm). The felt or other approved material shall be continuous to the top of walls and terminated at penetrations and building appendages in such a manner to meet the requirements of the exterior wall envelope as described in Section R703.1.

> **Exceptions:** ~~Such felt or material is permitted to be omitted~~ Omission of such water-resistive barrier is permitted in the following situations:
>
> 1. In detached accessory buildings.
> 2. ~~Under panel siding with shiplap joints or battens.~~
> ~~3.~~2. Under exterior wall finish materials as permitted in Table R703.4.
> ~~4.~~3. Under paperbacked stucco lath, when the paper backing is an approved weather-resistive sheathing paper.

CHANGE SIGNIFICANCE.   The changes within Section R703.2 and Table R703.8 are the results of several code change proposals processed both in the 2003/2004 code development cycle and the 2004/2005 code development cycle. While there are several clarifications and smaller changes, the most important change deals with the subject of protecting the exterior wall studs or sheathing with protective sheathing material in all exterior wall assemblies, regardless of the type of veneer or siding used. The Table R703.4 entry for "sheathing paper required" now shows "yes" for each and every siding or veneer type, meaning that regardless of the type and nature of siding, it is assumed that some moisture will penetrate the wall assembly and that protection for the exterior studs or sheathing must be provided by means of protective sheathing. There are some minor exceptions provided in the exceptions to Section R703.2 or within the table footnotes. The term *weather-resistive* has also

R703.2 and Table R703.4 continues

R703.2 and Table R703.4

Water-Resistive Barrier (R703.2) and Weather-Resistant Siding Attachment and Minimum Thickness (R703.4)

R703.2 and Table R703.4 continued

TABLE R703.4   Weather-Resistant Siding Attachment and Minimum Thickness

Siding Material		Nominal Thickness (inches)[a]	Joint Treatment	Weather Resistant Barrier ~~Sheathing Paper~~ Required
Horizontal aluminum[~~e~~]	Without insulation	0.019[f]	Lap	~~No~~ Yes
		0.024	Lap	~~No~~ Yes
	With insulation	0.019	Lap	~~No~~ Yes
Brick veneer[z] Concrete masonry veneer[z]		2 2	Section R703	Yes (Note ~~m~~l)
Hardboard[~~i~~k] Panel siding-vertical		7/16	~~Note g~~	~~See R703.2~~ Yes
Hardboard[~~i~~k] Lap-siding-horizontal		7/16	Note ~~r~~q	Yes
Steel[~~i~~h]		29 ga.	Lap	~~No~~ Yes

For SI: 1 inch = 25.4 mm.

a. Based on stud spacing of 16 inches on center where studs are spaced 24 inches, siding shall be applied to sheathing approved for that spacing.
b. Nail is a general description and shall be T-head, modified round head, or round head with smooth or deformed shanks.
c. Staples shall have a minimum crown width of 7/16-inch outside diameter and be manufactured of minimum 16 gage wire.
d. Nails or staples shall be aluminum, galvanized, or rust-preventative coated and shall be driven into the studs for fiberboard or gypsum backing.
e. Aluminum nails shall be used to attach aluminum siding.
f. Aluminum (0.019 inch) shall be unbacked only when the maximum panel width is 10 inches and the maximum flat area is 8 inches. The tolerance for aluminum siding shall be +0.002 inch of the nominal dimension.
~~g. If boards or panels are applied over sheathing or a weather resistant membrane, joints need not be treated. Otherwise, vertical joints shall occur at studs and be covered with battens or be lapped.~~
g. All attachments shall be coated with a corrosion-resistant coating.
h. Shall be of approved type.
i. Three-eighths-inch plywood shall not be applied directly to studs spaced more than 16 inches on center when long dimension is parallel to studs. Plywood 1/2-inch or thinner shall not be applied directly to studs spaced more than 24 inches on center. The stud spacing shall not exceed the panel span rating provided by the manufacturer unless the panels are installed with the face grain perpendicular to the studs or over sheathing approved for that stud spacing.
j. Wood board sidings applied vertically shall be nailed to horizontal nailing strips or blocking set 24 inches on center. Nails shall penetrate 1 1/2 inches into studs, studs and wood sheathing combined, or blocking. A weather-resistive membrane shall be installed weatherboard fashion under the vertical siding unless the siding boards are lapped or battens are used.

Type of Supports for the Siding Material and Fasteners[b,c,d]					
Wood or Wood Structural Panel Sheathing	**Fiberboard Sheathing into Stud**	**Gypsum Sheathing into Stud**	**Foam Plastic Sheathing into Stud**	**Direct to Studs**	**Number or Spacing of Fasteners**
0.120 nail 1½″ long	0.120 nail 2″ long	0.120 nail 2″ long	0.120 nail[x,y]	Not allowed	Same as stud spacing
0.120 nail 1½″ long	0.120 nail 2″ long	0.120 nail 2″ long	0.120 nail[x,y]	Not allowed	
0.120 nail 1½″ long	0.120 nail 2½″ long	0.120 nail 2½″ long	0.120 nail[y]	0.120 nail 1½″ long	
See Section R703 and Figure R703.7[h,g]					
Note o n	Note o n	Note o n	Note o n	Note o n	6″ panel edges 12″ inter. Sup.[n,o]
Note q p	Note q p	Note q p	Note q p	Note q p	Same as stud spacing 2 per bearing
0.113 nail 1¾″ Staple-1¾″	0.113 nail 2¾″ Staple-2½″	0.113 nail 2½″ Staple-2¼″	0.113 nail[x,y] Staple[x,y]	Not allowed	Same as stud spacing

k. Hardboard siding shall comply with AHA A 135.6.

l. For masonry veneer, a weather-resistive sheathing paper is not required over a sheathing that performs as a weather-resistive barrier when a 1-inch air space is provided between the veneer and the sheathing. When the 1-inch space is filled with mortar, a weather-resistive sheathing paper is required over studs or sheathing.

m. Vinyl siding shall comply with ASTM D 3679.

n. Minimum shank diameter of 0.092 inch, minimum head diameter of 0.225 inch, and nail length must accommodate sheathing and penetrate framing 1½ inches.

o. When used to resist shear forces, the spacing must be 4 inches at panel edges and 8 inches on interior supports.

p. Minimum shank diameter of ~~0.092~~ 0.099 inch, minimum head diameter of ~~0.225~~ 0.240 inch, and nail length must accommodate sheathing and penetrate framing 1½ inches.

q. Vertical end joints shall occur at studs and shall be covered with a joint cover or shall be caulked.

r. Fiber cement siding shall comply with the requirements of ASTM C 1186.

s. See Section R703.10.1.

t. Minimum 0.102″ smooth shank, 0.255″ round head.

u. Minimum 0.099″ smooth shank, 0.250″ round head.

v. See Section R703.10.2.

w. Face nailing: 2 nails at each stud. Concealed nailing: one 11 gage 1½ galv. roofing nail (0.371″ head diameter, 0.120″ shank) or 6d galv. box nail at each stud.

x. See Section R703.2 exceptions.

y. Minimum nail length must accommodate sheathing and penetrate framing 1½ inches.

z. Adhered masonry veneer shall comply with the requirements in Sections 6.1 and 6.3 of ACI 530/ASCE 5/TMS-402.

R703.2 and Table R703.4 continues

R703.2 and Table R703.4 continued

TABLE R703.4   Weather-Resistant Siding Attachment and Minimum Thickness

Siding Material	Nominal Thickness (inches)[a]	Joint Treatment	Weather Resistant Barrier ~~Sheathing Paper~~ Required
Stone veneer	2	Section R703	Yes (Note ~~m~~l)
Particleboard panels	$3/8$-$1/2$	~~Note g~~	~~Note g~~ Yes
	$5/8$	~~Note g~~	~~Note g~~ Yes
Plywood panel[i] (exterior grade)	$3/8$	~~Note g~~	~~Note g~~ Yes
Vinyl siding[m]	0.035	Lap	~~No~~ Yes
Wood[i] Rustic, drop	$3/8$ Min	Lap	~~No~~ Yes
Shiplap	$19/32$ Average	Lap	~~No~~ Yes
Bevel	$7/16$		
Butt tip	$3/16$	Lap	~~No~~ Yes
Fiber cement panel siding[r]	$5/16$	Note ~~t~~ s	Yes Note ~~y~~ x
Fiber cement lap siding[f]	$5/16$	Note ~~w~~ v	Yes Note ~~y~~ x

Type of Supports for the Siding Material and Fasteners[b,c,d]					
Wood or Wood Structural Panel Sheathing	Fiberboard Sheathing into Stud	Gypsum Sheathing into Stud	Foam Plastic Sheathing into Stud	Direct to Studs	Number or Spacing of Fasteners
See Section R703 and Figure R703.7[h,g]					
6d box nail (2″ × .099″)	6d box nail (2″ × .099″)	6d box nail (2″ × .099″)	box nail[e,y]	6d box nail, (2″ × .099″) ⁵⁄₈ not allowed	6″ panel edge 12″ inter. sup.
6d box nail (2″ × .099″)	8d box nail (2½″ × 0.113″)	8d box nail (2½″ × 0.113″)	box nail[e,y]	6d box nail (2″ × .099″)	
0.099 nail-2″	0.113 nail-2½″	0.099 nail-2″	0.113 nail[e,y]	0.099 nail-2″	6″ on panel edges 12″ inter. sup.
0.120 nail 1½″ Staple-1¾″	0.120 nail 2″ Staple-2½″	0.120 nail 2″ Staple-2½″	0.120 nail[e,y] Staple[e,y]	Not allowed	Same as stud spacing
Fastener penetration into stud-1″				0.113 nail-2½″ Staple-2″	Face nailing up to 6″ widths, 1 nail per bearing; 8″ widths and over, 2 nails per bearing
6d corrosion resistant nail[e,t]	6d corrosion resistant nail[e,t]	6d corrosion resistant nail[e,t]	6d corrosion resistant 2″ × 0.113″ nail[t,y]	4d corrosion resistant nail[e,u]	6″ oc on edges, 12″ oc on intermed. studs
6d corrosion resistant nail[e,t]	6d corrosion resistant nail[e,t]	6d corrosion resistant nail[e,t]	6d corrosion resistant 2″ × 0.113″ nail[t,y]	6d corrosion resistant nail[e,w]	Note * w

R703.2 and Table R703.4 continues

R703.2 and Table R703.4 continued

been changed in favor of *water resistive,* and a definition for water-resistive barrier has been provided in Section R202 to clarify the intended function of the material, which is protection against water. The use of sheathing paper to reduce water intrusion into buildings is believed necessary by many in the construction industry. Exterior claddings, although providing the first level of protection for water protrusion, are not designed to provide a perfect barrier to all water penetration. Therefore, a secondary line of defense is needed to protect moisture-sensitive building materials. The definition for the new term water-resistive barrier in the 2006 IRC is: A material behind an exterior wall covering that is intended to resist liquid water that has penetrated behind the exterior covering from further intruding into the exterior wall assembly.

CHANGE TYPE. Addition

CHANGE SUMMARY. New language is added, creating a specific method of protection for wood-based sheathing that is placed behind exterior plaster. This addition makes the IRC more consistent with the IBC for exterior plaster application.

2006 CODE: R703.6.3 Water-Resistive Barriers. Water-resistive barriers shall be installed as required in Section R703.2 and, where applied over wood-based sheathing, shall include a water-resistive vapor permeable barrier with a performance at least equivalent to two layers of Grade D paper.

> **Exception:** Where the water-resistive barrier that is applied over wood-based sheathing has a water resistance equal to or greater than that of 60 minute Grade D paper and is separated from the stucco by an intervening, substantially non-water-absorbing layer or designed drainage space.

CHANGE SIGNIFICANCE. The 2003 IRC provisions for the application of exterior plaster in Section 703.6 do not contain a specific method of protection for wood-based sheathing attached to the building exterior behind plaster. The weather-resistant protection provisions for building elements behind the exterior plaster are the same for all other exterior walls and are found in Section R703.2. The new language in the 2006 IRC requires that either two layers of Grade D paper or another vapor barrier equivalent to two layers of Grade D paper be applied to wood-based sheathing. Experience has shown that the typical method of protection for exterior wood sheathing has created problems for some types of exterior plaster. Grade D paper appears to have the proper permeability characteristics to prevent entrapment of

R703.6.3 continues

R703.6.3
Water-Resistive Barriers

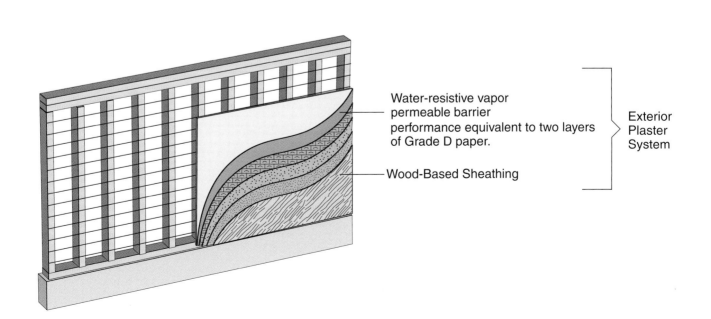

Water-resistive vapor permeable barrier performance equivalent to two layers of Grade D paper.

Wood-Based Sheathing

Exterior Plaster System

R703.6.3 continued

moisture, thereby eliminating most of such moisture-related problems behind plaster applications.

The exception addresses protection of wood-based sheathing for stucco systems specifically and provides yet another prescriptive method of exterior wood-based-sheathing protection when stucco exterior systems are used. The exception is merely intended to clarify the availability of alternative standard practices by recognizing stucco systems in which one of the two layers of weather-resistive barrier is replaced by a layer that, although not a Grade D paper, provides separation from the wet stucco application and provides a barrier to moisture from the stucco to the weather-resistive barrier.

R703.7
Stone and Masonry Veneer, General

CHANGE TYPE. Modification

CHANGE SUMMARY. The stone and masonry veneer section requirements and its extensive exceptions have been reorganized into two exceptions, two tables, and a figure for increased clarity and improved interpretation. Technical exceptions for the new Seismic Design Category D_0 have also been added.

2006 CODE: R703.7 Stone and Masonry Veneer, General. All stone and masonry veneer shall be installed in accordance with this chapter, Table R703.4 and Figure R703.7. Such veneers installed over a backing of wood or cold-formed steel shall be limited to the first story above grade and shall not exceed 5 inches (127 mm) in thickness.

Exceptions:

1. ~~In Seismic Design Categories A and B, exterior masonry veneer with a backing of wood or cold-formed steel framing shall not exceed 30 feet (9144 mm) in height above the noncombustible foundation, with an additional 8 feet (2348 mm) permitted for ends.~~

2. ~~In Seismic Design Category C, exterior masonry veneer with a backing of wood or cold-formed steel framing shall not exceed 30 feet (9144 mm) in height above the noncombustible foundation, with an additional 8 feet (2348 mm) permitted for gabled ends. In other than the topmost story, the length of~~

R703.7 continues

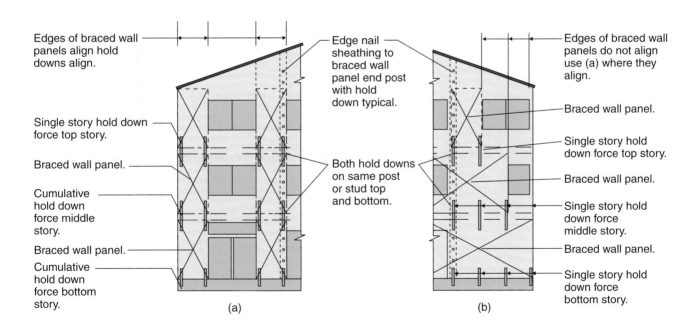

Figure R703.7(1) Hold Downs at Exterior and Interior Braced Wall Panels When Using Stone or Masonry Veneer.

(a) Braced wall panels stacked (aligned story to story). Use cumulative hold down force.
(b) Braced wall panels not stacked. Use single hold down force.

R703.7 continued

bracing shall be 1.5 times the length otherwise required in Chapter 6.

3. For detached one- or two- family dwellings with a maximum nominal thickness of 4 inches (102 mm) of exterior masonry veneer with a backing of wood frame located in Seismic Design Category D1, the masonry veneer shall not exceed 20 feet (6096 mm) in height above a noncombustible foundation, with an additional 8 feet (2438 mm) permitted for gabled ends, or 30 feet (9144 mm) in height with an additional 8 feet (2438 mm) permitted for gabled ends where the lower 10 feet (3048 mm) has a backing of concrete or masonry wall, provided the following criteria are met:

   3.1 Braced wall panels shall be constructed with a minimum of ⁷⁄₁₆-inch (11.1 mm) thick sheathing fastened with 8d common nails at 4 inches (102 mm) on center on panel edges and at 12 inches (305mm) on center on intermediate supports.

   3.2 The bracing of the top story shall be located at each end and at least every 25 feet (7620 mm) on center but not less than 45% of the braced wall line. The bracing of the first story shall be as provided in Table R602.10.1.

   3.3 Hold down connectors shall be provided at the ends of braced walls for the second floor to first floor wall assembly with an allowable design of 2100 lbs. (952.5 kg) Hold down connectors shall be provided at the ends of each wall segment of the braced walls for the first floor to foundation assembly with an allowable design of 3700 lbs. (1678 kg). In all cases, the hold down connector force shall be transferred to the foundation.

   3.4 Cripple walls shall not be permitted.

4. For detached one- and two-family dwellings with a maximum actual thickness of 3 inches (76 mm) of exterior masonry veneer with a backing of wood frame located in Seismic Design Category D2, the masonry veneer shall not exceed 20 feet (6096 mm) in height above a noncombustible foundation, with an additional 8 feet (2438 mm) permitted for gabled ends, or 30 feet (9144 mm) in height with an additional 8 feet (2438 mm) permitted for gabled ends where the lower 10 feet (3048 mm) has a backing of concrete or masonry wall, provided the following criteria are met:

   4.1 Braced wall panels shall be constructed with a minimum of ⁷⁄₁₆-inch (11.1 mm) thick sheathing fastened with 8d common nails at 4 inches (102 mm) on center on panel edges and at 12 inches (305mm) on center on intermediate supports.

   4.2 The bracing of the top story shall be located at each end and at least every 25 feet (7620 mm) on center but not less than 55% of the braced wall line. The bracing of the first story shall be as provided in Table R602.10.1.

   4.3 Hold down connectors shall be provided at the ends of braced walls for the second floor to first floor wall assem-

~~bly with an allowable design of 2300 lbs.(1043 kg). Hold down connectors shall be provided at the ends of each wall segment of the braced walls for the first floor to foundation assembly with an allowable design of 3900 lbs. (1769 kg). In all cases, the hold down connector force shall be transferred to the foundation.~~

4.4 ~~Cripple walls shall not be permitted.~~

1. For all buildings in Seismic Design Categories A, B, and C, exterior stone or masonry veneer, as specified in Table R703.7(1), with a backing of wood or steel framing shall be permitted to the height specified in Table R703.7(1) above a noncombustible foundation. Wall bracing at exterior and interior braced wall lines shall be in accordance with Sections R602.10 or R603.7, and the additional requirements of Table R703.7(1).

2. For detached one- or two-family dwellings in Seismic Design Categories D_0, D_1, and D_2, exterior stone or masonry veneer, as specified in Table R703.7(2), with a backing of wood framing shall be permitted to the height specified in Table R703.7(2)

R703.7 continues

TABLE R703.7(1) **Stone or Masonry Veneer Limitations and Requirements, Wood or Steel Framing, Seismic Design Categories A, B, and C**

Seismic Design Category	Number of Wood or Steel Famed Stories	Maximum Height of Veneer Above Noncombustible Foundation (feet)[a]	Maximum Nominal Thickness of Veneer (inches)	Maximum Weight of Veneer (psf)[b]	Wood or Steel Framed Story	Minimum Sheathing Amount (percent of braced wall line length)[c]
A or B	steel: 1 or 2 wood: 1, 2 or 3	30	5	50	all	Table R602.10.1 or Table R603.7
C	1	30	5	50	1 only	Table R602.10.1 or Table R603.7
	2	30	5	50	top	Table R602.10.1 or Table R603.7
					bottom	1.5 times length required by Table R602.10.1 or 1.5 times length required by Table R603.7
	Wood only: 3	30	5	50	top	Table R602.10.1
					middle	1.5 times length required by Table R602.10.1
					bottom	1.5 times length required by Table R602.10.1

For SI: 1 inch = 25.4 mm, 1 foot = 304.8 mm, 1 pound per square foot = 0.479 KPa.

a. An additional 8 feet (2438 mm) is permitted for gable end walls. See also story height limitations of Section R301.3.
b. Maximum weight is installed weight and includes weight of mortar, grout, lath, and other materials used for installation. Where veneer is placed on both faces of a wall, the combined weight shall not exceed that specified in this table.
c. Applies to exterior and interior braced wall lines.

R703.7 continued

above a noncombustible foundation. Wall bracing and hold downs at exterior and interior braced wall lines shall be in accordance with Sections R602.10 and R602.11 and the additional requirements of Table R703.7(2). In Seismic Design Categories D_0, D_1, and D_2, cripple walls shall not be permitted, and required interior braced wall lines shall be supported on continuous foundations.

CHANGE SIGNIFICANCE. Section 703.7, stone and masonry veneer, in the 2003 code contains four exceptions for various Seismic Design Categories (SDCs), with Exceptions 3 and 4 each containing four subsections of their own. Following the addition of a new SDC, designated as D_0 in the 2004 supplement, three additional exceptions related to the new SDC were proposed in code change RB184-04/05. The overall result of these changes was an extensive number of exceptions with their own subsections that created confusion and made application of the code difficult. To address this problem, the public comment submitted for the ICC's Final Action Agenda reformatted all of the exceptions into two exceptions plus two tables and a figure to improve the usability of the code.

The new technical provisions for the SDC D_0 contain prescriptive language for the methods of support and backing to construct such veneers to certain heights.

TABLE R703.7(2) Stone or Masonry Veneer Limitations, One- and Two-Family Detached Dwellings, Wood Framing, Seismic Design Categories D_0, D_1, and D_2

Seismic Design Category	Number of Wood Framed Stories[a]	Maximum Height of Veneer Above Noncombustible Foundation or Foundation Wall (feet)	Maximum Nominal Thickness of Veneer (inches)	Maximum Weight of Veneer (psf)[b]	Wood Frames Story	Minimum Sheathing Amount (percent of braced wall line length)[c]	Minimum Sheathing Thickness and Fastening	Single Story Hold Down Force (lbs)[d]	Cumulative Hold Down Force (lbs)[e]
D_0	1	20[f]	4	40	1 only	35%	$7/16$-inch wood structural panel sheathing with 8d common nails spaced at 4 inches on center at panel edges. 12 inches on center at intermediate supports. 8d common nails at 4 inches on center at braced wall panel end posts with hold down attached.	N/A	—
	2	20[f]	4	40	top	35%		1900	—
					bottom	45%		3200	5100
	3	30[g]	4	40	top	40%		1900	—
					middle	45%		3500	5400
					bottom	60%		3500	8900
D_1	1	20[f]	4	40	1 only	45%		2100	—
	2	20[f]	4	40	top	45%		2100	—
					bottom	45%		3700	5800
	3	20[f]	4	40	top	45%		2100	—
					middle	45%		3700	5800
					bottom	60%		3700	9500
D_2	1	20[f]	3	30	1 only	55%		2300	—
	2	20[f]	2	30	top	55%		2300	—
					bottom	55%		3900	6200

NA = Not applicable. For SI: 1 inch = 25.4 mm. 1 foot = 304.8 mm. 1 pound per square foot = 0.479 KPa, 1 pound force = 4.448 N.

a. Cripple walls are not permitted in Seismic Design Categories D_0, D_1, and D_2.

b. Maximum weight is installed weight and includes weight of mortar, grout, and lath, and other materials used for installation.

c. Applies to exterior and interior braced wall lines.

d. Hold down force is minimum allowable stress design load for connector providing uplift tie from wall framing at end of braced wall panel at the story below, or to foundation or foundation wall. Use single story hold down force where edges of braced wall panels do not align; a continuous load path to the foundation shall be maintained. (See Figure R.703.7(1)[b])

e. Where hold down connectors from stories above align with stories below, use cumulative hold down force to size middle and bottom story hold down connectors. (See Figure R.703.7(1)[a])

f. The veneer shall not exceed 20 feet (6096 mm) in height above a noncombustible foundation, with an additional 8 feet (2438 mm) permitted for gable end walls, or 30 feet (9144 mm) in height with an additional 8 feet (2438 mm) for gable end walls where the lower 10 feet (3048 mm) has a backing of concrete or masonry wall. See also story height limitations of Section R301.3.

g. The veneer shall not exceed 30 feet (9144 mm) in height above a noncombustible Foundation, with an additional 8 feet (2438 mm) permitted for gable end walls. See also story height limitations of Section R301.3.

R703.8

Flashing

CHANGE TYPE.   Deletion

CHANGE SUMMARY.   The provisions for self-flashing windows have been eliminated.

2006 CODE: R703.8 Flashing.   Approved corrosion-resistant flashing shall be ~~provided~~ applied shingle fashion ~~in the exterior wall envelope~~ in such a manner as to prevent entry of water into the wall cavity or penetration of water to the building structural framing components. The flashing shall extend to the surface of the exterior wall finish. ~~and shall be installed to prevent water from reentering the exterior wall envelope~~ Approved corrosion-resistant flashings shall be installed at all of the following locations:

1. ~~At top of all e~~Exterior window and door openings. ~~in such a manner as to be leak-proof, except that self-flashing windows having a continuous lap of not less than 1-⅛ inches (28mm) over the sheathing material around the perimeter of the opening, including corners, do not require additional flashing; jamb flashing may also be omitted when specifically approved by the building official.~~ Flashing at exterior window and door

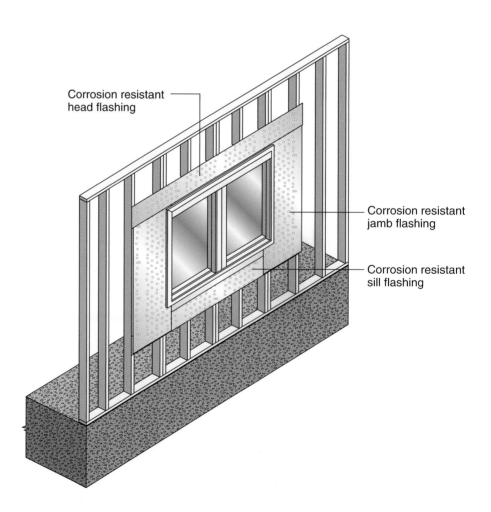

Corrosion resistant head flashing

Corrosion resistant jamb flashing

Corrosion resistant sill flashing

<u>openings shall extend to the surface of the exterior wall finish or to the water-resistive barrier for subsequent drainage.</u>

2. At the intersection of chimneys or other masonry construction with frame or stucco walls, with projecting lips on both sides under stucco copings.

3. Under and at the ends of masonry, wood or metal copings and sills.

4. Continuously above all projecting wood trim.

5. Where exterior porches, decks, or stairs attach to a wall or floor assembly of wood-frame construction.

6. At wall and roof intersections.

7. At built-in gutters.

CHANGE SIGNIFICANCE.   Minor rewording and relocation of certain phrases have improved the clarity of the flashing requirements. A reference to flashing being extended to the water-resistive barrier for exterior windows and doors is intended to update the code to reflect common practice and national standards such as ASTM E2112. The most significant change in this section is related to the elimination of self-flashing windows. The 2003 code allows elimination of flashing where ". . . self-flashing windows having a continuous lap of not less than 1⅛ inches (28 mm) over the sheathing material around the perimeter of the opening, including corners, do not require additional flashing." This allowance for self-flashing windows has been eliminated because window flanges are intended for mounting, not flashing, purposes. ASTM E2112 in Section 8.1.1.1 clearly requires flashing to be integrated with window flanges. The phrase "jamb flashing may also be omitted when specifically approved by the building official" has also been eliminated, so flashing is now automatically required at all exterior window jambs. The improvements in the 2006 code provisions for flashing are intended to better address the protection of the building envelope from exterior water penetration, which is a common source of problems at exterior windows and doors.

R802.1.5

Structural Log Members

CHANGE TYPE. Addition

CHANGE SUMMARY. New grading rules and a referenced standard have been added for roof-ceiling, wall, and floor construction of structural log members.

2006 CODE: R802.1.5 Structural Log Members. Stress grading of structural log members of nonrectangular shape, as typically used in log buildings shall be in accordance with ASTM D3957. Such structural log members shall be identified by the grade mark of an approved lumber grading or inspection agency. In lieu of a grade mark on the material, a certificate of inspection as to species and grade issued by a lumber-grading or inspection agency meeting the requirements of this section shall be permitted to be accepted.

CHANGE SIGNIFICANCE. Logs used as structural members for floors, walls, and roofs are common in certain parts of the country. The 2003 code does not contain any provisions or references for such members and how they should be graded or regulated. There are some ICC-ES evaluation reports for some log manufacturers under some legacy model building codes. The new 2006 code text provides the grading rules and makes reference to ASTM D3957 for nonrectangular logs. The same exact language has been repeated in Section R802.1.5 for roof-ceiling log frame members, Section R602.1.3 for wall log members, and Section R502.1.6 for floor log frame members.

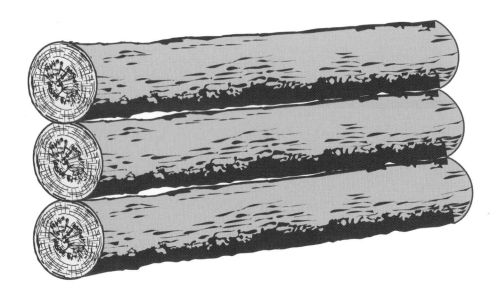

R802.3.1 and Table R802.5.1(9)
Ceiling Joist and Rafter Connections

CHANGE TYPE.   Addition

CHANGE SUMMARY.   New language has been added to provide prescriptive provisions for the connection of ceiling joists to rafter ties, and a modification table has been added to Table R802.5.1(9) to increase the number of connectors when the ties are located higher than the top of the wall plates.

2006 CODE: R802.3.1 Ceiling Joist and Rafter Connections. Ceiling joists and rafters shall be nailed to each other in accordance with Tables ~~R602.3(1) and~~ R802.5.1(9), and the ~~assembly~~ rafter shall be nailed to the top wall plate in accordance with Table R602.3(1). Ceiling joists shall be continuous or securely joined in accordance with Table 802.5.1(9) where they meet over interior partitions and nailed to adjacent rafters to provide a continuous tie across the building when such joists are parallel to the rafters.

Where ceiling joists are not connected to the rafters at the top wall plate, joists connected higher in the attic shall be installed as rafter ties, or rafter ties shall be installed to provide a continuous tie. Where ceiling joists are not parallel to rafters, rafter ties shall be installed. Rafter ties shall be a minimum of 2-inch by 4-inch (51 mm by 102 mm) (nominal), installed in accordance with ~~subflooring or metal straps attached at the ends of the rafters shall be installed in a manner to provide a continuous tie across the building, or rafter shall be tied to 1 inch by 4 inch (25.4 mm by 102 mm) (nominal) minimum size crossties. The~~ the connection~~s~~ requirements in Table R802.5.1(9), ~~shall be in accordance with Table R602.3(1)~~ or connections of equivalent capacities shall be provided. Where ceiling joists or rafter ties are not provided ~~at the top plate~~, the ridge formed by these rafters shall be supported by a wall or girder designed in accordance with accepted engineering practice.

Collar ties or ridge straps to resist wind uplift shall be connected in the upper third of the attic space, in accordance with Table R602.3(1). ~~Rafter~~ Collar ties shall be a minimum of 1-inch by 4-inch

R802.3.1 and Table R802.5.1(9) continues

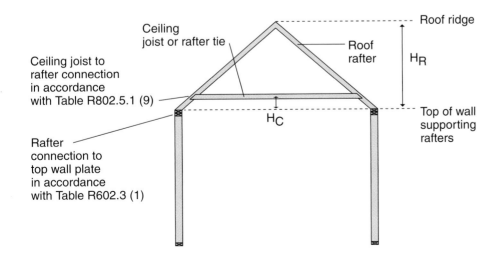

Ceiling joist or rafter tie

Roof ridge

Roof rafter

H_R

Ceiling joist to rafter connection in accordance with Table R802.5.1 (9)

H_C

Top of wall supporting rafters

Rafter connection to top wall plate in accordance with Table R602.3 (1)

R802.3.1 and Table R802.5.1(9)
continued

(25 mm by 102 mm) (nominal), spaced not more than 4 feet (1219 mm) on center.

CHANGE SIGNIFICANCE.   Section R802.3.1 contains provisions for the connection of ceiling joists to rafters. Ceiling joists that are parallel to rafters are tied to the rafters to resist the outward push of the rafters. When ceiling joists are perpendicular to rafters and cannot be tied to them, rafter ties must be provided to resist this outward push. This section, in the 2003 code, assumes that the ceilings or rafter ties are being connected to the rafters at the top of the bearing wall. Many designs in conventional construction practices, however, place the ceiling joists or the rafter ties higher up in the attic. Cross-ties or ceilings placed higher cannot resist as much load as when they are placed lower, and for this reason Tables R802.5.1(1) through (9) apply an adjustment factor for the allowable length of rafters. The last table in these series, Table R802.5.1(9), which provides the minimum connection requirements, in the 2003 code is inconsistent with the previous eight tables because it does not consider the effects of rafter ties or ceilings being placed higher in the attic. The 2006 changes add new language within Section R802.3.1 and add a new footnote, "g," to Table R802.5.1(9), similar to the same footnote in the previous eight tables, to account for the possibility of ties or ceiling joists being located at a point higher than the top of the bearing wall. The application of the new footnote will result in additional connectors being required on the basis of the distance that ceiling joists or rafter ties are located above the top plate of the wall.

TABLE R802.5.1(9)   Rafter/Ceiling Joist Heel Joint Connections[a,b,c,d,e,f,g]

Rafter Slope	Rafter Spacing (inches)	Ground Snow Load (psf)											
		30 psf				50 psf				70 psf			
		Roof Span (feet)											
		12	20	28	36	12	20	28	36	12	20	28	36
		Required Number of 15d Common Nails[a,b] per Heel Joint Splices[c,d,e,f]											
3:12	12	4	6	8	11	5	8	12	15	6	11	15	20
	16	5	8	11	14	6	11	15	20	8	14	20	26
	24	7	11	16	21	9	16	23	30	12	21	30	39
4:12	12	3	5	6	8	4	6	9	11	5	8	12	15
	16	4	6	8	11	5	8	12	15	6	11	15	20
	24	5	9	12	16	7	12	17	22	9	16	23	29
5:12	12	3	4	5	7	3	5	7	9	4	7	9	12
	16	3	5	7	9	4	7	9	12	5	9	12	16
	24	4	7	10	13	6	10	14	18	7	13	18	23
7:12	12	3	3	4	5	3	4	5	7	3	5	7	9
	16	3	4	5	6	3	5	7	9	4	6	9	11
	24	3	5	7	9	4	7	10	13	5	9	13	17
9:12	12	3	3	3	4	3	3	4	5	3	4	5	7
	16	3	3	4	5	3	4	5	7	3	5	7	9
	24	3	4	6	7	3	6	8	10	4	7	10	13
12:12	12	3	3	3	3	3	3	3	4	3	3	4	5
	16	3	3	3	4	3	3	4	5	3	4	5	7
	24	3	3	4	6	3	4	6	8	3	6	8	10

For SI: 1 inch = 25.4 mm, 1 foot = 304.8 mm, 1 pound per square foot = 0.0479 KPa.

a. 40d box nails shall be permitted to be substituted for 16d common nails.
b. Nailing requirements shall be permitted to be reduced 25 percent if nails are clinched.
c. Heel joint connections are not required when the ridge is supported by a load-bearing wall, header, or ridge beam.
d. When intermediate support of the rafter is provided by vertical struts or purlins to a loadbearing wall, the tabulated heel joint connection requirements shall be permitted to be reduced proportionally to the reduction in span.
e. Equivalent nailing patterns are required for ceiling joist to ceiling joist lap splices.
f. When rafter ties are substituted for ceiling joists, the heel joint connection requirement shall be taken as the tabulated heel joint connection requirement for two-thirds of the actual rafter-slope.
g. Tabulated heel joint connection requirements assume that ceiling joists or rafter ties are located at the bottom of the attic space. When ceiling joists or rafter ties are located higher in the attic, heel joint connection requirements shall be increased by the following factors:

H_c/H_R	Heel Joint Connection Adjustment Factor
1/3	1.5
1/4	1.33
1/5	1.25
1/6	1.2
1/10 or less	1.11

where: H_c = Height of ceiling joists or rafter ties measured vertically above the top of the rafter support walls.
H_R = Height of roof ridge measured vertically above the top of the rafter support walls.

Table R802.5.1(1) through Table R802.5.1(8)

Rafter Spans for Common Lumber Species

CHANGE TYPE. Modification

CHANGE SUMMARY. Ceiling joists that are connected to rafters and act as rafter ties, or the rafter ties used to resists the outward thrust forces of the rafters are no longer allowed to be placed higher than the lower one-third of the attic height.

2006 CODE: **Table R802.5.1(2) Rafter Spans for Common Lumber Species (Roof live load = 20 psf, ceiling attached to rafters, L/Δ = 240)** (see pages 167–168).

CHANGE SIGNIFICANCE. Rafter ties and ceiling joists that are connected to rafters and act as rafter ties are most effective when they are placed nearest to the heel of the rafters. The outward roof-rafter-thrust forces resulting from roof loads are the highest at the heel, and the longer the span of the rafter, the higher such thrust loads are. It is for this reason that the code assumes rafter ties or the ceiling joists are connected to roof rafters at the top of the bearing wall and closest to the rafter heel. Current practice, however, on many occasions places these critical load-resisting elements not at the top of the wall but higher in the attic area. Although this practice allows for higher ceilings, it creates a potential problem for the exterior walls (which might be kicked outward) and the roof structure, if the outward thrust forces are not adequately resisted by the rafter ties or the ceiling joists. To deal with this issue, rafter span tables contain a footnote "a" that requires the length of the rafter be multiplied by a length-reduction factor when the ties are not placed at the top plate of the wall. The higher the ties are located in the attic, the smaller the length-reduction factor-multiplier is, resulting in smaller rafter spans.

The footnote length-reduction factors in the 2003 code allow the rafters to be placed higher in the attic without any limitation. It appears the 2003 code allows such ties to be located all the way at the

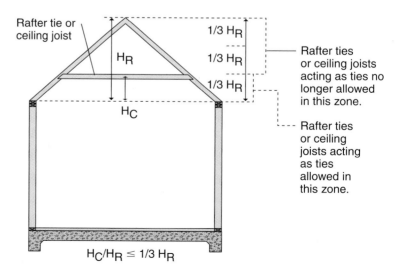

TABLE R802.5.1(2)
Rafter Spans for Common Lumber Species (Roof live load = 20 psf, ceiling attached to rafters, L/Δ = 240)

Rafter Spacing (inches)	Species and Grade		Dead Load = 10 psf					Dead Load = 20 psf				
			2 × 4	2 × 6	2 × 8	2 × 10	2 × 12	2 × 4	2 × 6	2 × 8	2 × 10	2 × 12
			Maximum Rafter Spans[a]									
			(feet-inches)	(feet-inches)	(feet-inches)	(feet-inches)	(feet-inches)	(feet-inches)	(feet-inches)	(feet-inches)	(feet-inches)	(feet-inches)
12	Douglas fir-larch	SS	10-5	16-4	21-7	Note b	Note b	10-5	16-4	21-7	Note b	Note b
	Douglas fir-larch	#1	10-0	15-9	20-10	Note b	Note b	10-0	15-4	19-5	23-9	Note b
	Douglas fir-larch	#2	9-10	15-6	20-5	25-8	Note b	9-10	14-4	18-2	22-3	25-9
	Douglas fir-larch	#3	8-7	12-6	15-10	19-5	22-6	7-5	10-10	13-9	16-9	19-6
	Hem-fir	SS	9-10	15-6	20-5	Note b	Note b	9-10	15-6	20-5	Note b	Note b
	Hem-fir	#1	9-8	15-2	19-11	25-5	Note b	9-8	14-11	18-11	23-2	Note b
	Hem-fir	#2	9-2	14-5	19-0	24-3	Note b	9-2	14-2	17-11	21-11	25-5
	Hem-fir	#3	8-7	12-6	15-10	19-5	22-6	7-5	10-10	13-9	16-9	19-6
	Southern pine	SS	10-3	16-1	21-2	Note b	Note b	10-3	16-1	21-2	Note b	Note b
	Southern pine	#1	10-0	15-9	20-10	Note b	Note b	10-0	15-9	20-10	25-10	Note b
	Southern pine	#2	9-10	15-6	20-5	Note b	Note b	9-10	15-1	19-5	23-2	Note b
	Southern pine	#3	9-1	13-6	17-2	20-3	24-1	7-11	11-8	14-10	17-6	20-11
	Spruce-pine-fir	SS	9-8	15-2	19-11	25-5	Note b	9-8	15-2	19-11	25-5	Note b
	Spruce-pine-fir	#1	9-5	14-9	19-6	24-10	Note b	9-5	14-4	18-2	22-3	25-9
	Spruce-pine-fir	#2	9-5	14-9	19-6	24-10	Note b	9-5	14-4	18-2	22-3	25-9
	Spruce-pine-fir	#3	8-7	12-6	15-10	19-5	22-6	7-5	10-10	13-9	16-9	19-6
16	Douglas fir-larch	SS	9-6	14-11	19-7	25-0	Note b	9-6	14-11	19-7	24-9	Note b
	Douglas fir-larch	#1	9-1	14-4	18-11	23-9	Note b	9-1	13-3	16-10	20-7	23-10
	Douglas fir-larch	#2	8-11	14-1	18-2	22-3	25-9	8-6	12-5	15-9	19-3	22-4
	Douglas fir-larch	#3	7-5	10-10	13-9	16-9	19-6	6-5	9-5	11-11	14-6	16-10
	Hem-fir	SS	8-11	14-1	18-6	23-8	Note b	8-11	14-1	18-6	23-8	Note b
	Hem-fir	#1	8-9	13-9	18-1	23-1	Note b	8-9	12-11	16-5	20-0	23-3
	Hem-fir	#2	8-4	13-1	17-3	21-11	25-5	8-4	12-3	15-6	18-11	22-0
	Hem-fir	#3	7-5	10-10	13-9	16-9	19-6	6-5	9-5	11-11	14-6	16-10
	Southern pine	SS	9-4	14-7	19-3	24-7	Note b	9-4	14-7	19-3	24-7	Note b
	Southern pine	#1	9-1	14-4	18-11	24-1	Note b	9-1	14-4	18-10	22-4	Note b
	Southern pine	#2	8-11	14-1	18-6	23-2	Note b	8-11	13-0	16-10	20-1	23-7
	Southern pine	#3	7-11	11-8	14-10	17-6	20-11	6-10	10-1	12-10	15-2	18-1
	Spruce-pine-fir	SS	8-9	13-9	18-1	23-1	Note b	8-9	13-9	18-1	23-0	Note b
	Spruce-pine-fir	#1	8-7	13-5	17-9	22-3	25-9	8-6	12-5	15-9	19-3	22-4
	Spruce-pine-fir	#2	8-7	13-5	17-9	22-3	25-9	8-6	12-5	15-9	19-3	22-4
	Spruce-pine-fir	#3	7-5	10-10	13-9	16-9	19-6	6-5	9-5	11-11	14-6	16-10
19.2	Douglas fir-larch	SS	8-11	14-0	18-5	23-7	Note b	8-11	14-0	18-5	22-7	Note b
	Douglas fir-larch	#1	8-7	13-6	17-9	21-8	25-2	8-4	12-2	15-4	18-9	21-9
	Douglas fir-larch	#2	8-5	13-1	16-7	20-3	23-6	7-9	11-4	14-4	17-7	20-4
	Douglas fir-larch	#3	6-9	9-11	12-7	15-4	17-9	5-10	8-7	10-10	13-3	15-5
	Hem-fir	SS	8-5	13-3	17-5	22-3	Note b	8-5	13-3	17-5	22-3	25-9
	Hem-fir	#1	8-3	12-11	17-1	21-1	24-6	8-1	11-10	15-0	18-4	21-3
	Hem-fir	#2	7-10	12-4	16-3	20-0	23-2	7-8	11-2	14-2	17-4	20-1
	Hem-fir	#3	6-9	9-11	12-7	15-4	17-9	5-10	8-7	10-10	13-3	15-5
	Southern pine	SS	8-9	13-9	18-1	23-1	Note b	8-9	13-9	18-1	23-1	Note b
	Southern pine	#1	8-7	13-6	17-9	22-8	Note b	8-7	13-6	17-2	20-5	24-4
	Southern pine	#2	8-5	13-3	17-5	21-2	24-10	8-4	11-11	15-4	18-4	21-6
	Southern pine	#3	7-3	10-8	13-7	16-0	19-1	6-3	9-3	11-9	13-10	16-6
	Spruce-pine-fir	SS	8-3	12-11	17-1	21-9	Note b	8-3	12-11	17-1	21-0	24-4
	Spruce-pine-fir	#1	8-1	12-8	16-7	20-3	23-6	7-9	11-4	14-4	17-7	20-4
	Spruce-pine-fir	#2	8-1	12-8	16-7	20-3	23-6	7-9	11-4	14-4	17-7	20-4
	Spruce-pine-fir	#3	6-9	9-11	12-7	15-4	17-9	5-10	8-7	10-10	13-3	15-5

Table R802.5.1(1) through Table R802.5.1(8) continues

Table R802.5.1(1) through Table
    R802.5.1(8) continued

TABLE R802.5.1(2)
Rafter Spans for Common Lumber Species (Roof live load = 20 psf, ceiling attached to rafters, L/Δ = 240)

Rafter Spacing (inches)	Species and Grade		Dead Load = 10 psf					Dead Load = 20 psf				
			2 × 4	2 × 6	2 × 8	2 × 10	2 × 12	2 × 4	2 × 6	2 × 8	2 × 10	2 × 12
			Maximum Rafter Spans[a]									
			(feet-inches)	(feet-inches)	(feet-inches)	(feet-inches)	(feet-inches)	(feet-inches)	(feet-inches)	(feet-inches)	(feet-inches)	(feet-inches)
24	Douglas fir-larch	SS	8-3	13-0	17-2	21-10	Note b	8-3	13-0	16-7	20-3	23-5
	Douglas fir-larch	#1	8-0	12-6	15-10	19-5	22-6	7-5	10-10	13-9	16-9	19-6
	Douglas fir-larch	#2	7-10	11-9	14-10	18-2	21-0	6-11	10-2	12-10	15-8	18-3
	Douglas fir-larch	#3	6-1	8-10	11-3	13-8	15-11	5-3	7-8	9-9	11-10	13-9
	Hem-fir	SS	7-10	12-3	16-2	20-8	25-1	7-10	12-3	16-2	19-10	23-0
	Hem-fir	#1	7-8	12-0	15-6	18-11	21-11	7-3	10-7	13-5	16-4	19-0
	Hem-fir	#2	7-3	11-5	14-8	17-10	20-9	6-10	10-0	12-8	15-6	17-11
	Hem-fir	#3	6-1	8-10	11-3	13-8	15-11	5-3	7-8	9-9	11-10	13-9
	Southern pine	SS	8-1	12-9	16-10	21-6	Note b	8-1	12-9	16-10	21-6	Note b
	Southern pine	#1	8-0	12-6	16-6	21-1	25-2	8-0	12-3	15-4	18-3	21-9
	Southern pine	#2	7-10	12-3	15-10	18-11	22-2	7-5	10-8	13-9	16-5	19-3
	Southern pine	#3	6-5	9-6	22-1	14-4	17-1	5-7	8-3	10-6	12-5	24-9
	Spruce-pine-fir	SS	7-8	12-0	25-10	20-2	24-7	7-8	12-0	15-4	18-9	21-9
	Spruce-pine-fir	#1	7-6	11-9	14-10	18-2	21-0	6-11	10-2	12-10	15-8	18-3
	Spruce-pine-fir	#2	7-6	11-9	14-10	18-2	21-0	6-11	10-2	12-10	15-8	18-3
	Spruce-pine-fir	#3	6-1	8-10	11-3	13-8	15-11	5-3	7-8	9-9	11-10	13-9

Check sources for availability of lumber in lengths greater than 20 feet.
For SI: 1 inch = 25.4 mm, 1 foot = 304.8 mm, 1 pound per square foot = 0.0475 KPa.

a. The tabulated rafter spans assume that ceiling joists are located at the bottom of the attic space or that some other method of resisting the outward push of the rafters on the bearing walls, such as rafter ties, is provided at that location. When ceiling joists or rafter ties are located higher in the attic space, the rafter spans shall be multiplied by the factors given below:

H_c/H_R	Rafter Span Adjustment Factor
~~2/3 or greater~~	~~0.50~~
~~1/2~~	~~0.58~~
1/3	0.67
1/4	0.76
1/5	0.83
1/6	0.90
1/7.5 or less	1.00

where: H_c = Height of ceiling joists or rafter ties measured vertically above the top of the rafter support walls.
       H_R = Height of roof ridge measured vertically above the top of the rafter support walls.

top of the attic with a length adjustment factor of 0.5. Experience has shown that ceiling joists or rafter ties located too close to the top of the rafters lose their effectiveness significantly, and the lateral deflection of the rafters below the ties are so excessive that engineering calculations become necessary. For this reason, the 2006 code has eliminated the allowance that rafter ties and ceilings can be placed higher than the lower one-third of the attic area.

R802.10.2.1
Applicability Limits Wood Truss Design

CHANGE TYPE. Addition

CHANGE SUMMARY. A new section, R802.10.2.1, outlining the limitations of Section R802.10.2, has been added, creating applicability limits for wood truss design similar to applicability limits for cold-formed-steel roof framing design.

2006 CODE: R802.10.2.1 Applicability Limits. The provisions of this section shall control the design of truss roof framing when snow controls for buildings not greater than 60 feet (18 288 mm) in length perpendicular to the joist, rafter or truss span, not greater than 36 feet (10 973 mm) in width parallel to the joist span or truss, not greater than two stories in height with each story not greater than 10 feet (3048 mm) high, and roof slopes not smaller than 3:12 (25% slope) or greater than 12:12 (100% slope). Truss roof framing constructed in accordance with the provisions of this section shall be limited to sites subjected to a maximum design wind speed of 110 miles per hour (49 m/s), Exposure A, B, or C and a maximum ground snow load of 70 psf (3352 Pa). Roof snow load is to be computed as: $0.7p_g$.

CHANGE SIGNIFICANCE. Wood trusses are required to be designed "in accordance with accepted engineering practice." The bracing requirements for wood trusses and the design and manufacture of metal-

R802.10.2.1 continues

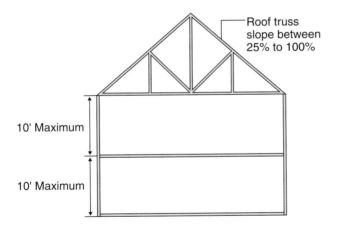

Roof truss slope between 25% to 100%

10' Maximum

10' Maximum

Maximum design wind speed of 110 mph, exposure A, B, or C and maximum ground snow load of 70 psf.

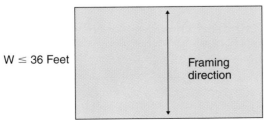

W ≤ 36 Feet

Framing direction

L ≤ 60 Feet

R802.10.2.1 continued

plate-connected wood trusses are in part referenced to national standards such as TPI/HIB (Truss Plate Institute/Handling, Installing, and Bracing guidelines) and ANSI/TPI 1 *(National Design Standard for Metal-Plate-Connected Wood Truss Construction)* (TPI/HIB; the HIB-91 "booklet," *Commentary & Recommendations for Handling, Installing & Bracing Metal Plate Connected Wood Trusses,* has been replaced by *BCSI 1-03: Guide to Good Practice For Handling, Installing & Bracing of Metal Plate Connected Wood Trusses).* None of these provisions have changed in the 2006 code. What *has* changed is a new section that inserts the limitations of these provisions, very similar to the limitations found in Section R804.1.1 for cold-formed-metal roof members, where snow controls the roof design, because where and how snow falls and how it applies load to a structure are independent of structure type and structural framing used. These limitations are applicable for roof trusses when snow controls the design and the building dimensions are 36 feet by 60 feet and two stories in height. The wind-load limitations are set at 110 mph in the lower seismic design categories, and the maximum ground snow-load limitation is 70 psf.

CHANGE TYPE.   Addition

CHANGE SUMMARY.   New provisions for conditioned attic spaces have been introduced for increased energy efficiency and to accommodate recent trends in improved residential construction.

2006 CODE: R806.4 Conditioned Attic Assemblies: Unvented conditioned attic assemblies (spaces between the ceiling joists of the top story and the roof rafters) are permitted under the following conditions:

1. No interior vapor retarders are installed on the ceiling side (attic floor) of the unvented attic assembly.

2. An air-impermeable insulation is applied in direct contact to the underside/interior of the structural roof deck. "Air-impermeable" shall be defined by ASTM E 283.

   Exception: In zones 2B and 3B, insulation is not required to be air impermeable.

3. In the warm humid locations as defined in N1101.2.1:

   3.1 For asphalt roofing shingles: A 1 perm (5.7×10^{-11} kg/s. m^2. Pa) or less vapor retarder (determined using Procedure B of ASTM E 96) is placed to the exterior of the structural roof deck; that is, just above the roof structural sheathing.

   3.2 For wood shingles and shakes: a minimum continuous $\frac{1}{4}$-inch (6 mm) vented air space separates the shingles/shakes and the roofing felt placed over the structural sheathing.

4. In zones 3 through 8 as defined in N1101.2 sufficient insulation is installed to maintain the monthly average temperature

R806.4 continues

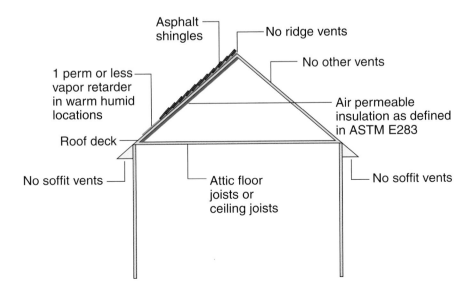

R806.4 continued

of the condensing surface above 45° F (7° C). The condensing surface is defined as either the structural roof deck or the interior surface of an air-impermeable insulation applied in direct contact to the underside/interior of the structural roof deck. "Air-impermeable" is quantitatively defined by ASTM E 283. For calculation purposes, an interior temperature of 68° F (20° C) is assumed. The exterior temperature is assumed to be the monthly average outside temperature.

CHANGE SIGNIFICANCE. The conditioned attic assembly provisions found in 2006 code Section R806.4 are similar in concept to the new Section R408.3, Unvented Crawl Space, introduced in Chapter 4. In fact, these were both part of the same code change proposal within code change EC48-2003/2004, which was intended to simplify the code in relation to energy-efficiency provisions. It has become a common practice in recent years for HVAC units to be placed in the attic areas of residential buildings. The ventilation requirements of attics in the 2003 code do not allow for conditioned attic spaces. This has been a major source of energy inefficiency because attic installed HVAC units and their ducts, a master source of air leakage, are within an environment extremely hot in summer and extremely cold in winter. Additionally, attic ventilation and the introduction of exterior air within the building envelope in highly humid regions appear to be an unsound practice that has potential implications for the integrity of building elements. In recent years, many advocates of building sciences and energy efficiency have researched and experimented in construction of new homes with the concept of conditioned attic spaces. It appears the results of all such research and experimentation have been positive and reflect the correctness of the theories for conditioned attic spaces. As such, the 2006 code has incorporated new prescriptive provisions for conditioned attic assemblies which employ different requirements based on the climatic region where the building is located.

CHANGE TYPE.   Addition

CHANGE SUMMARY.   Hail exposure categories for roof coverings have been introduced through new definitions and a new figure.

2006 CODE: R903.5 Hail Exposure.   Hail exposure, as specified in Sections R903.5.1 and R903.5.2, shall be determined using Figure R903.5.

R903.5.1 Moderate Hail Exposure.   One or more hail days with hail diameters larger than 1.5 in (38 mm) in a twenty (20) year period.

R903.5.2 Severe Hail Exposure.   One or more hail days with hail diameters larger than or equal to 2.0 in (50 mm) in a twenty (20) year period.

CHANGE SIGNIFICANCE.   A recent study conducted by the Institute for Business and Home Safety (IBHS) has shown that various types of "non-impact-resistant" roof coverings impacted by substantial hail needed repair or replacement. For example, for asphalt shingles, the study showed that approximately 45% of the non-impact-resistant asphalt shingle roofs needed repair or replacement after being struck by hailstones with diameters between 1.0 inch and 2.0 inches. It should

R903.5, R903.5.1, R903.5.2, and Figure R903.5 continues

R903.5, R903.5.1, R903.5.2, and Figure R903.5

Hail Exposure (R903.5), Moderate Hail Exposure (R903.5.1), Severe Hail Exposure (R903.5.2), Hail Exposure Map (Figure R903.5)

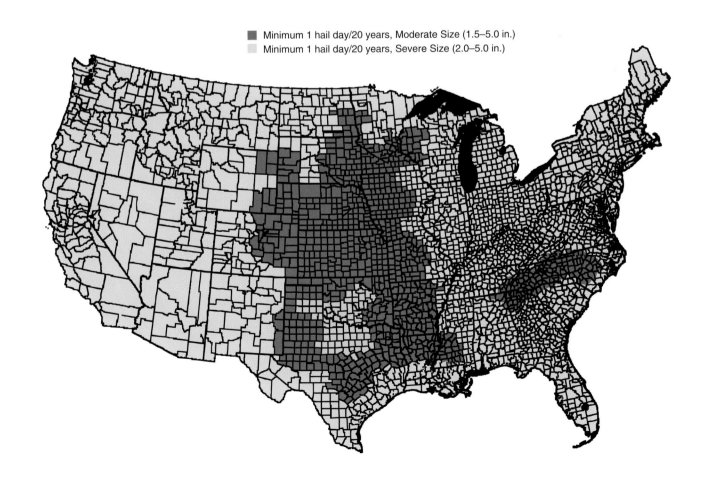

■ Minimum 1 hail day/20 years, Moderate Size (1.5–5.0 in.)
▪ Minimum 1 hail day/20 years, Severe Size (2.0–5.0 in.)

*R903.5, R903.5.1, R903.5.2,
and Figure R903.5 continued*

also be noted that even though the lifespan of an asphalt shingle roof covering varies with respect to product line and environmental conditions, most such roofs are warranted (excluding damage from hail) for 20 years or more. Throughout large portions of the Great Plains and Southeastern United States, hailstorms producing hail with diameters of 1.5 inches or larger are expected at mean recurrence intervals of 20 years or less. Thus, in such regions, damaging hail is expected within a period of time less than or equal to the lifetime of most roof coverings, including asphalt shingles. Because the 2003 IRC does not require the consideration of impact resistance in the selection of roof-covering materials, an attempt was made through a code change proposal to bring such a requirement into the 2006 code. However, many portions of the proposed code change were eliminated and were not incorporated into the 2006 code. The portions eliminated would have required asphalt, clay, concrete, and metal roof coverings to be impact-resistant. What remained of the proposed code change and is now in the 2006 code are definitions for hail exposure categories and the hail exposure map that appear in Section R903.5 and an asphalt shingle replacement provision that appears as new item 4 in Section R907.3. As such, the 2006 code still does not require impact resistance in roof coverings. The original code change proposal would have required impact resistance for roof coverings through Underwriters Laboratory Standard UL 2218 and Factory Mutual Standard FM 4473. These sections were ultimately not included in the 2006 code because most updated revisions to these standards were not available.

The new hail exposure map shown in the figure was developed with statistical modeling software, with use of actual hail data from the National Climatic Data Center.

CHANGE TYPE. Addition

CHANGE SUMMARY. The wrappers for asphalt shingles being used in areas subject to wind speeds of 110 mph or greater must now be labeled to indicate that the shingles use special fastening methods complying with ASTM D 3161, Class F.

2006 CODE: R905.2.6 Attachment. Asphalt shingles shall have the minimum number of fasteners required by the manufacturer. For normal application, asphalt shingles shall be secured to the roof with not less than four fasteners per strip shingle or two fasteners per individual shingle. Where the roof slope exceeds 20 units vertical in 12 units horizontal (167% slope), special methods of fastening are required. For roofs located where the basic wind speed per Figure 301.2(4) is 110 mph (49 m/s) or higher, special methods of fastening are required. Special fastening methods shall be tested in accordance with ASTM D 3161, ~~modified to use a wind speed of 110 mph (177 km/h)~~ <u>Class F. Asphalt shingle wrappers shall bear a label indicating compliance with ASTM D 3161 Class F.</u>

~~Shingles classified using ASTM D 3161 are acceptable for use in wind zones less than 110 mph (49 m/s). Shingles classified using ASTM D 3161 Class F are acceptable for use in all cases where special fastening is required.~~

CHANGE SIGNIFICANCE. There are currently no requirements in the 2003 code for labeling of shingle wrappers containing shingles that are intended for use in high-wind areas. In regions with basic wind speeds of 110 mph or greater, the building officials, consumers, roofers, and suppliers have difficulty verifying that shingle products being used are in fact approved for special fastening methods in accordance with ASTM D3161, Class F. The new language in the 2006 code requires that all such shingle wrappers be labeled to indicate special fastening methods in accordance with ASTM D3161, Class F.

R905.2.6 continues

R905.2.6 continued This will eliminate confusion and misapplication of shingles in high-wind regions. ASTM Standard D 3161 designates as Class F shingles that are tested to a wind speed of 110 mph. The last paragraph, which provided clarification about where shingles tested to ASTM D 3161 could be used, has been deleted from this section and has been relocated, without any technical changes, to Section R905.2.4.1. This has been done for immediate clarification of this issue up front, in the section on asphalt shingles. The latest version of ASTM D 3161 is the 2003 edition, which is now referenced in the 2006 code, whereas the 2003 code referenced the ASTM D 3161-99.

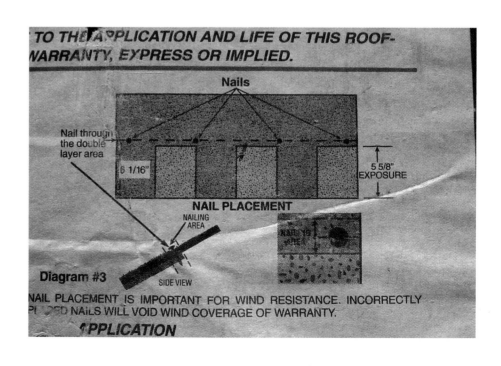

R905.2.7.1

Ice Barrier

CHANGE TYPE. Modification

CHANGE SUMMARY. The criteria and the trigger for requiring "ice barriers" have been modified from "average daily temperature of 25° F or less" in the 2003 code to whether "there has been a history of ice forming along the eaves . . ." in the 2006 code.

2006 CODE: R905.2.7.1 Ice ~~Protection~~ Barrier. In areas where ~~the average daily temperature in January is 25° F (−4° C) or less or when~~ there has been a history of ice forming along the eaves causing a backup of water as designated in Table R301.2(1) ~~criteria so designates~~, an ice barrier that consists of at least two layers of underlayment cemented together or of a self-adhering polymer modified bitumen sheet, shall be used in lieu of normal underlayment and extend from the ~~eave's edge~~ lowest edges of all roof surfaces to a point at least 24 inches (610 mm) inside the exterior wall line of the building.

> **Exception:** Detached accessory structures that contain no conditioned floor area.

CHANGE SIGNIFICANCE. Requiring an ice barrier on the basis of "areas where the average daily temperature in January [is] 25° F (−4° C) or less" is almost impossible. The following map shows the average January temperature of 25° or less for the year 2004. Similar annual maps are available for several decades prior to 2004, and such maps

R905.2.7.1 continues

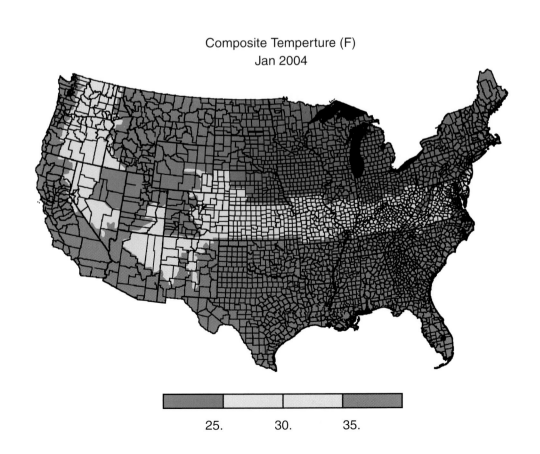

Composite Temperture (F)
Jan 2004

25.   30.   35.

R905.2.7.1 continued

from 1970 to 2004 were reviewed by the proponents of the code change that resulted in the new 2006 language. The review revealed that during this 54-year period, Januaries in which the average temperature was 25° or less were extremely variable. For example, the only January in the past 45 years during which this was the average temperature in any part of the State of Tennessee was in 1977. Also, the northwest part of Nevada had this average temperature only in 1993. At no other times were these and many other areas ever in this average temperature range. In comparison, January 1990 is shown as very warm in most parts of the United States. An even more critical criterion is which years are chosen to determine the average. For example, in southwestern Ohio there is a history of ice damming, even though this area had this average temperature for only 8 of the past 35 years. Using local history as the criterion for the ice barrier requirement is shown in Table R301.2(1) footnote "i." It is therefore more logical to make the ice barrier determination on the basis of local historic data and experience and not a temperature range average, which may vary over many decades. The maps mentioned here are from the National Oceanic and Atmospheric Administration (NOAA)-CIRES Climate Diagnostics Center, and the dataset information was obtained from the National Climatic Data Center. The maps can be accessed on the NOAA Web site, http://www.cdc.noaa.gov/USclimate/USclimdivs.html. The criteria entered to produce these maps were as follows: variable, "temperature"; plot of the "mean" for month of "January to January"; "range of years," the same year; low, "25" degrees; high, "35" degrees; and contour interval, "5." Other coordinating changes with this concept have been made in the Table R301.2 (1) column heading and footnote "i" and in Sections R905.4.3, R905.5.3, R905.6.3, R905.7.3, and R905.8.3.

Another modification in this section of the 2006 code is that the term *eave's edge* has been changed to "lowest edges of all roof surfaces" because of the misapplication of this provision. It is the lowest edges of a roof that, because of gravity, melting snow, and ice, will refreeze and cause a back-up of ice. The new terminology eliminates some confusion and misapplication due to various definitions for "eave" and other roof elements such as "rake." In its application, it is the lowest edge of the roof where the underlayment is installed.

CHANGE TYPE. Modification

CHANGE SUMMARY. The metal roof covering standards table has been modified to include additional metal roofing materials that have been successfully used in the field and a new table has been created to separate and clarify the corrosion resistance standards from the material standards.

2006 CODE:

TABLE R905.10.3(1) Metal Roof Coverings Standards

Roof Covering Type	Standard Application Rate/Thickness
Galvanized steel	ASTM A 653 G-90 zinc coated, 0.013 inch thick minimum
Stainless steel	ASTM A 240, 300 Series Alloys
Steel	ASTM A 924
Lead-coated copper	ASTM B 101
Cold rolled copper	ASTM B 370 minimum 16 oz/square ft and 12 oz/square ft high yield copper for metal-sheet roof-covering systems; 12 oz/square ft for preformed metal shingle systems. ~~CDA A115~~
Hard lead	2 lb/sq ft
Soft lead	3 lb/sq ft
Aluminum	ASTM B 209, 0.024 minimum thickness for rollformed panels and 0.019 inch minimum thickness for pressformed shingles.
Terne (tin) and terne-coated stainless	Terne coating of 40 lb per double base box, field painted where applicable in accordance with manufacturer's installation instructions.
Zinc	0.027 inch minimum thickness: 99.995% electrolytic high grade zinc with alloy additives of copper (0.08–0.20%), titanium (0.07%–0.12%) and aluminum (0.015%).
~~Aluminum zinc alloy-coated steel~~	~~ASTM A 792 AZ 50~~
~~Prepainted Steel~~	~~ASTM A 755~~

For SI: 1 ounce per square foot = 0.305 kg/m^2, 1 pound per square foot = 4.214 kg/m^2, 1 inch = 25.4 mm, 1 pound = 0.454 kg.

CHANGE SIGNIFICANCE. Section R905.10.3 requires that metal-sheet roof coverings installed over structural decking comply with Table R905.10.3. The 2003 code Table R905.10.3 identifies nine different metal roof coverings and indicates the applicable standard for each type. As such, other metal roof coverings not listed in this table are not recognized automatically by the IRC. The 2006 code table has added some new entries such as aluminum alloy-coated steel, corre-

Table R905.10.3 continues

Table R905.10.3 continued

TABLE R905.10.3(2)   Minimum Corrosion Resistance

55% Aluminum-zinc alloy coated steel	ASTM A 792 AZ 50
5% Aluminum alloy-coated steel	ASTM A 875 GF 60
Aluminum-coated steel	ASTM A 463 T2 65
Galvanized steel	ASTM A 653 G-90
Prepainted steel	ASTM A 755[a]

a. Paint systems in accordance with ASTM A 755 shall be applied over steel products with corrosion-resistant coatings complying with ASTM A 792, ASTM A 875, ASTM A 463, or ASTM A 653.

sponding to ASTM A875, for aluminum-coated steel, corresponding to ASTM A463, and stainless steel. The 2003 IRC Table 905.10.3 contains a mix of information related to material standards and corrosion resistance standards. For further clarity, in the 2006 code, the corrosion resistance standards have been separated and placed in a new Table R905.10.3(2). The aluminum-coated steel listed in Table R905.10.3(2) is the one with the coating designation of T2 65. This product is specified by its coating designation and consists of designations T1 25, T1 40, T1 100, T2 LC, T2 65, and T2 100, each of which has a different weight of coating (oz/ft^2). Similarly, the 5% aluminum alloy-coated steel specified in the table is the one with the coating designation GF60, which has minimum requirements for a coating weight of 0.60 oz/ft^2 total for both sides, 0.20 oz/ft^2 for one side on the basis of a triple spot test, and 0.50 oz/ft^2 total for both sides on the basis of a single spot test. This ASTM standard, A875, contains a Table X2.1 that converts the coating weight to an approximate coating-thickness equivalent.

In the entry for copper, in Table R905.10.3(1), the reference to "CDA 4115" has been deleted. This change removes the entire reference to *Copper in Architecture-Design Handbook* from the IRC. This document does not meet the strict compliance requirements for it to be con-

sidered a reference standard as outlined by the ICC policy; because it is not developed through a consensus process, and is not a testing, material properties, design, or installation standard in the true sense of the words. The removal of this reference does not leave the user of the code without a resource, as there is a reference standard that addresses the material properties contained within Table R905.10.3(1). This is consistent with the other entries of the table.

R905.12.2, R905.13.2, and R905.15.2

Material Standards

CHANGE TYPE.   Addition and Deletion

CHANGE SUMMARY.   The outdated referenced standards RMA RP-1, RP-2, and RP-3 for thermoset single-ply roof covering have been deleted, and two new referenced standards, ASTM D6754 for thermoplastic single-ply roof coverings and ASTM D6694 for liquid-applied roof coatings, have been added.

2006 CODE: R905.12.2 Material Standards.   Thermoset single-ply roof coverings shall comply with ~~RMA RP-1, RP-2, or RP-3, or~~ ASTM D 4637, ASTM D 5019, or CGSB 37-GP-52M.

R905.13.2 Material Standards.   Thermoplastic single-ply roof coverings shall comply with ASTM D4434, <u>ASTM D6754</u>, or CSGB 37-GP-54M.

R905.15.2 Material Standards.   Liquid-applied roof coatings shall comply with ASTM C 836, C 957, D 1227, D 3468, D 6083, or <u>D 6694</u>.

CHANGE SIGNIFICANCE.   The three referenced standards, Rubber Manufacturers Association (RMA) RP-1-90 (*Minimum Requirements for Non-Reinforced Black EPDM* [Ethylene Propylene Diene Monomer] *Rubber Sheets*), RP-2-90 *(Minimum Requirements for Fabric-Reinforced Black EPDM Rubber Sheets),* and RP-3-85 *(Minimum Requirements for Fabric-Reinforced Black Polychloroprene Rubber Sheets),* have been deleted without substitution because the technical committee that developed and maintained the referenced RMA standards is no longer active. The standards are not being maintained and published and are therefore not available for use. ASTM D 4637-96, *Specification for EPDM Sheet Used in Single-Ply Roof Membrane,* which is currently ref-

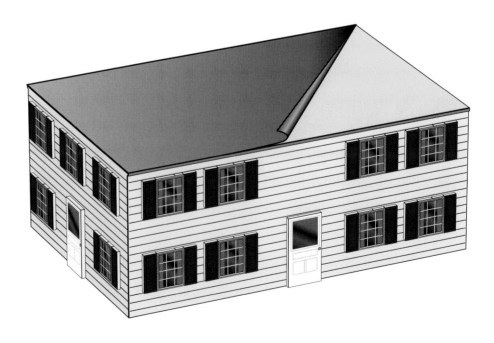

erenced in the IRC, was developed from the technical requirements of RP-1, RP-2, and RP-3 with the intent of replacing the three outdated RMA standards with a more comprehensive and current standard. The referenced ASTM standard contains the technical requirements necessary to achieve compliance with the minimum standards mandated by the code.

The new standard referenced in Section R905.13.2 is for a commonly used thermoplastic single-ply roof covering called ketone ethylene ester–based sheet and must comply with ASTM D6754. The other new standard referenced in Section R905.15.2 is ASTM D6694, which allows an alternative silicone-based elastomeric coating as a protective coating for spray polyurethane foam insulation. ASTM D6694-01 is titled *Standard Specification for Liquid-Applied Silicone Coating Used in Spray Polyurethane Foam Roofing.*

R907.3

Recovering Versus Replacement

CHANGE TYPE. Clarification

CHANGE SUMMARY. For the replacement or recovering of existing roof coverings, a new Exception 3 has been added in the 2006 code, clarifying that recoating of existing spray polyurethane foam roofing systems is allowed.

2006 CODE: R907.3 Recovering Versus Replacement. New roof coverings shall not be installed without first removing existing roof coverings where any of the following conditions occur:

1. Where the existing roof or roof covering is water-soaked or has deteriorated to the point that the existing roof or roof covering is not adequate as a base for additional roofing.
2. Where the existing roof covering is wood shake, slate, clay, cement or asbestos-cement tile.
3. Where the existing roof has two or more applications of any type of roof covering.
4. For asphalt shingles, when the building is located in an area subject to moderate or severe hail exposure according to Figure R903.5

Exceptions:
1. Complete and separate roofing systems, such as standing-seam metal roof systems, that are designed to transmit the roof loads directly to the building's structural system and that do not rely on existing roofs and roof coverings for support shall not require the removal of existing roof coverings.
2. Metal panel, metal shingle, and concrete and clay tile roof coverings shall be permitted to be installed over existing wood shake roofs when applied in accordance with Section R907.4.

3. The application of new protective coating over existing spray polyurethane foam roofing systems shall be permitted without tear-off of existing roof coverings.

CHANGE SIGNIFICANCE.   New roof coverings are not allowed to be placed over existing roof coverings unless certain conditions are present. The code in general intends that multiple applications of roof coverings would not add unforeseen dead loads and that other unsanitary or unsafe conditions be removed and new roof coverings applied. The 2003 code contains two exceptions that allow recovering existing roofs. Exception 1 deals with separate roofing systems that transmit the roof loads directly to the building structure, and Exception 2 deals with metal, concrete, and clay tiles installed in accordance with Section R907.4, Roof Recovering. Under the 2003 code provisions, it is not clear whether recoating of existing spray-polyurethane-foam roofing systems is allowed or prohibited. Since recoating can add many years to the life of an existing spray-polyurethane-foam roofing system without adding significant weight to the roof assembly or compromising the long-term performance of the roofing assembly, a new Exception 3 has been added in the 2006 code, clarifying that such recoating is allowed. The industry practices for the recoating of an existing spray-polyurethane-foam roofing system are provided in ASTM D-6705-01, *Standard Guide for the Repair and Recoat of Spray Polyurethane Foam Roofing Systems,* which is now referenced in the new Exception 3.

R1003.15

Flue Area (Masonry Fireplace)

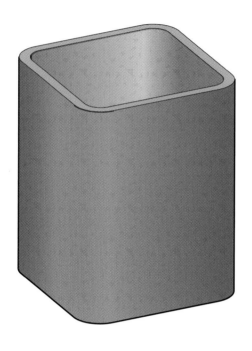

CHANGE TYPE. Clarification and Modification

CHANGE SUMMARY. Flue size determination for masonry fireplaces is based on either Option 1, found in Section R1003.15.1, or Option 2, found in Section R1003.15.2. This change takes the language related to the cross-sectional area of clay flue linings found in Option 2 and repeats it in Option 1, so it is clear the cross-sectional area of clay flue lining is determined by the same method, regardless of which option is used. The 2003 code Table R1001.11(2), which is numbered as R1003.14(2) in the 2006 code, has been modified to give more flue size options.

2006 CODE: ~~R1001.12~~ R1003.15 Flue Area (Masonry Fireplace). Flue sizing for chimneys serving fireplaces shall be in accordance with Section ~~R1001.12.1~~ R1003.15.1 or Section ~~R1001.12.2~~ R1003.15.2.

~~R1001.12.1~~ R1003.15.1 Option 1. Round chimney flues shall have a minimum net cross-sectional area of at least $\frac{1}{12}$ of the fireplace opening. Square chimney flues shall have a minimum net cross-sectional area of $\frac{1}{10}$ of the fireplace opening. Rectangular chimney flues with an aspect ratio less than 2 to 1 shall have a minimum net cross-sectional area of $\frac{1}{10}$ of the fireplace opening. Rectangular chimney flues with an aspect ratio of 2 to 1 or more shall have a minimum net cross sectional area of $\frac{1}{8}$ of the fireplace opening. Cross sectional areas of clay flue linings are shown in Tables R1001.14(1) and R1001.14(2) or as provided by the manufacturer or as measured in the field. The height of the chimney shall be measured from the firebox floor to the top of the chimney flue.

~~R1001.12.2~~ R1003.15.2 Option 2. The minimum net cross-sectional area of the chimney flue shall be determined in accordance with Figure R1003.15.2. A flue size providing at least the equivalent net cross-sectional area shall be used. Cross sectional areas of clay flue linings are provided in Tables R1001.14(1) and R1001.14(2) or as provided by the manufacturer or as measured in the field. The height of the chimney shall be measured from the firebox floor to the top of the chimney flue.

CHANGE SIGNIFICANCE. The determination of cross-sectional area of clay flue linings has been somewhat confusing when Option 1 is chosen. Because the 2003 code, in Option 1, does not specify what methodology is to be used for determining the clay flue lining cross-section, many believe that the methodology of the fraction of the fireplace opening ($\frac{1}{12}$, $\frac{1}{10}$, or $\frac{1}{8}$ of fireplace opening) is also to be used for clay flue linings. This conclusion is derived from the fact that Option 2 clearly and separately provides Tables R1003.14(1) and R1003.14(2) or the manufacturer's methodology for clay flue linings. The change in the 2006 code takes the exact same language found in Option 2 for clay flue linings and repeats it in Option 1, so that there is no doubt that the procedure for clay flue linings is the same, regardless of which option is selected. Clay flue linings are manufactured according to ASTM C315 and are manufactured from fire clay, shale, surface clay, or a combination of these materials that when formed and fired to suitable temperatures, yields a product that is strong and durable. Table R1003.14.2 revisions are intended to provide better corresponding flue size outside dimensions to cross sectional area. For example, a 20″ × 20″ flue requires a 286 square inch cross sectional area, but the same size flue in the 2006 code requires a 298 square inch cross sectional area.

Chapter 11

Energy Efficiency

It is generally believed by most code users that the provisions contained within the International Energy Conservation Code (IECC) and Chapter 11 of the IRC are difficult to understand, difficult to enforce, and too complex. Despite numerous sources of support, many of which are available at no charge through the U.S. Department of Energy (DOE), such anxiety and the complexities of the energy provisions have persisted. Inconsistency in application of the code provisions, inconsistency in the requirements in their application to different buildings, and the confusion about and misapplication of the provisions result in a lower level of compliance. On the basis of years of study, workshops, and feedback from the code users, the DOE prepared and submitted a comprehensive-code-change proposal during the first 18-month code change cycle (2003/2004 cycle). The proposed code change was assigned number EC48-03/04 for ICC Code Development Process, was ultimately approved as modified, and resulted in a complete rewriting of IRC Chapter 11 and much of the IECC. There were additional changes made to various sections of chapter 11 during the second 18-month code change cycle (2004/2005 cycle). These changes are intended to make Chapter 11 simpler, easier to understand and remember, and consistent for buildings within a jurisdiction or neighboring jurisdictions that form a larger community. As such, code users will find that the entire 2003 code Chapter 11 has been deleted and replaced with a new chapter 11. The new 2006 code Chapter 11 is intended to be technically consistent and at approximately the same level of stringency in comparison with the 2003 code. Because a reprint of the entire Chapter 11 is not the focus of this book, five selected sections of more pronounced significance have been presented. For a complete text of this chapter, refer to the 2006 IRC.

N1101.2
Compliance

CHANGE TYPE. Modification

CHANGE SUMMARY. It is made explicit that using either the IECC or Chapter 11 of the IRC is acceptable for compliance with the energy-efficiency provisions, and a new figure, N1101.2, has been introduced for the new climate zones.

2006 CODE: **N1101.2 Compliance.** Compliance shall be demonstrated by either meeting the requirement of the *International Energy Conservation Code* or meeting the requirements of this chapter. Climate zones from Figure N1101.2 or Table N1101.2 shall be used in determining the applicable requirements from this chapter.

CHANGE SIGNIFICANCE. It has now been made explicitly clear that the IRC allows either Chapter 11 of the IRC or the provisions of the IECC to be used to comply with the IRC energy requirements. The code user may now evaluate both options and use the one that fits the project best, as these two methods may result in different requirements. This new section also introduces a new figure, N1101.2, which shows the various climate zones in the country. The 2003 code's original 19 climate zones based on Heating Degree Days (HDD) have been combined, reorganized, and changed into eight climate zones based on multiple climate variables, so that both heating and cooling considerations are accommodated. Zones 1 through 7 apply to various parts of the

N1101.2 continues

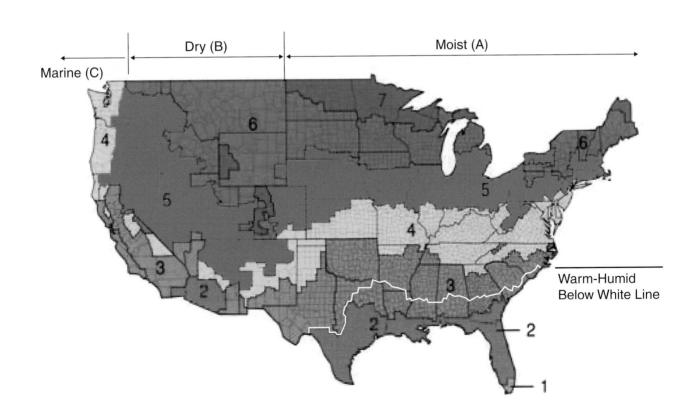

N1101.2 continued United States and are defined by county lines. Zones 7 and 8 apply to various parts of Alaska, and Hawaii is Zone 1. The climate zone table of the 2003 code has been replaced with the new table N1101.2 to reflect the new climate zones. The climate zones have been divided into marine, dry, and moist to deal with levels of humidity. For more details and background on the development of the new climate zones, code users can refer to the white paper at the following site: http://www.energycodes.gov/implement/pdfs/climate_paper_review_ draft_rev.pdf.

CHANGE TYPE. Modification

CHANGE SUMMARY. New marking methodology for sprayed polyurethane foam (SPF) has been introduced because the typical method of installing markers throughout the attic does not work for this sprayed-type insulation.

2006 CODE: N1101.4 Building Thermal Envelope Insulation. An *R*-value identification mark shall be applied by the manufacturer to each piece of building thermal envelope insulation 12 inches (305 mm) or more wide. Alternately, the insulation installers shall provide a certification listing the type, manufacturer and *R*-value of insulation installed in each element of the building thermal envelope. For blown or sprayed insulation (fiberglass and cellulose), the initial installed thickness, settled thickness, settled *R*-value, installed density, coverage area and number of bags installed shall be listed on the certification. For sprayed polyurethane foam (SPF) insulation, the installed thickness of the area covered and R-value of installed thickness shall be listed on the certification. The insulation installer shall sign, date and post the certification in a conspicuous location on the job site.

N1101.4.1 Blown or Sprayed Roof/Ceiling Insulation. The thickness of blown-in or sprayed roof/ceiling insulation (fiberglass or cellulose) shall be written in inches (mm) on markers that are installed at least one for every 300 ft² (28 m²) throughout the attic space. The markers shall be affixed to the trusses or joists and marked with the minimum initial installed thickness with numbers a minimum of 1 inch (25 mm) in height. Each marker shall face the attic access opening. Spray polyurethane foam thickness and installed R-value shall be listed on certification provided by the insulation installer.

CHANGE SIGNIFICANCE. The marking requirements for the purposes of inspections and documentation of the certificate for the job site are similar to the requirements found in 2003 code Section N1101.3.1, with the exception of a format change for further clarity

N1101.4 and N1101.4.1 continues

N1101.4 and N1101.4.1

Building Thermal Envelope Insulation (N1101.4) and Blown or Sprayed Roof/Ceiling Insulation (N1101.4.1)

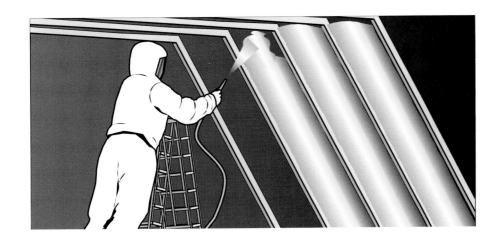

N1101.4 and N1101.4.1 continued

and providing a separate identification method for SPF. Measuring blown-in insulation such as fiberglass or cellulose by attaching markers to the joist or truss addresses the settling characteristic of blown-in insulation and therefore its settled R-value. Markers also provide a convenient method to inspect an entire attic from a single vantage point such as the attic access opening. Even though SPF is similar to blown-in insulation, it has completely different physical properties. One such difference is in the settlement characteristics. SPF does not settle like blown-in insulation; it actually expands upon initial application and covers any initial thickness marker. Another difference involves adhesive characteristics. SPF has tremendous adhesive properties, and when it is applied in uniform thickness over joists/trusses or roof/ceilings, it produces a wavelike pattern of highs and lows in attics, reflective of the cavity between joists/trusses. This wavelike pattern does not lend itself easily to visual inspection from a single vantage point within an attic. SPF is typically and most efficiently inspected by randomly probing the foam at visually apparent low spots. After installation, SPF is typically covered with an ignition barrier. This would also prevent markers from providing meaningful measurements of thickness.

N1101.8
Certificate

CHANGE TYPE.   Addition

CHANGE SUMMARY.   A new requirement has been added for an energy-efficiency certificate to be posted inside the electrical panel.

2006 CODE: **N1101.8 Certificate.**   A permanent certificate shall be posted inside the electrical distribution panel. The certificate shall be completed by the builder or registered design professional. The certificate shall list the predominant R-values of insulation installed in or on ceiling/roof, walls, foundation (slab, basement wall, crawlspace wall and/or floor) and ducts outside conditioned spaces; U-factors for fenestration; and the solar heat gain coefficient (SHGC) of fenestration. Where there is more than one value for each component, the certificate shall list the value covering the largest area. The certificate shall list the type and efficiency of heating, cooling, and service water heating equipment.

CHANGE SIGNIFICANCE.   This new section is intended to increase consumers' awareness of the energy-efficiency ratings for various

N1101.8 continues

N1101.8 continued building elements by requiring the builder or the registered design professional to complete a certificate and place it inside the electrical panel. The certificate must disclose the building's R-values, U-factors, and HVAC efficiencies. The energy efficiency of a building as a system is a function of many elements considered as parts of a whole, and it is almost always impossible to have a proper identification and analysis of a building's energy efficiency once the building is completed and many elements are not readily accessible. This information is also very valuable for existing buildings undergoing alterations or additions and for sizing mechanical equipment.

CHANGE TYPE. Modification

CHANGE SUMMARY. The window size and percentage thresholds for the applicability of the prescriptive provisions of the IRC energy provisions have been deleted, and a new methodology based on Table N1102.1 has been established.

2006 CODE: N1102.1 Insulation and Fenestration Criteria. The building thermal envelope shall meet the requirements of Table N1102.1 based on the climate zone specified in Table N1101.2.

N1102.1.3 Total UA Alternative. If the total building thermal envelope UA (sum of *U*-factor times assembly area) is less than or equal to the total UA resulting from using the *U*-factors in Table N1102.1.2 (multiplied by the same assembly area as in the proposed building), the building shall be considered in compliance with Table N1102.1. The UA calculation shall be done using a method consistent with the ASHRAE *Handbook of Fundamentals* and shall include the thermal bridging effects of framing materials. The SHGC requirements shall be met in addition to UA compliance.

CHANGE SIGNIFICANCE. The prescriptive envelope requirements are no longer dependent upon the percentage of window area. The 2003 code in Sections N1101.2.1, N1101.2.2, and N1102.1 establishes the thresholds of 15% glazing area for detached one- and two-family dwellings and 25% glazing area for townhouses for the IRC prescriptive provisions to be used. Such buildings with glazing areas greater than 15% and 25%, respectively, are referred to IECC Chapters 4 and 5. This threshold has now been deleted and no longer exists in the 2006 code. The building envelope thermal requirements are now determined according to the new Table N1102.1. Either IRC chapter 11 or the IECC could now be used for residential buildings, regardless of window sizes. For further details on the basis and analysis for elimination of window-size thresholds, visit this site: http://www.energycodes.gov/implement/pdfs/wwr_elimination.pdf

The total UA alternative methodology allows for comparing the entire building thermal envelope UA value to the sum of individual U factors from Table N1102.1.2 and if the total UA value is the same as or less than the total UA from Table N1102.1.2, then the building is considered in compliance. The calculation must be done by a method consistent with the ASHRAE Handbook referenced in the text of the code. Since the UA requirements vary by building component/assembly, it is important that the same area for each component/assembly be used for both the base case and the proposed building. To illustrate the application of this provision, if the proposed house has 400 square feet of window area, then the UA for this component/assembly for purposes of the UA alternative would be 400 square feet times the prescriptive window requirement. This value would be added together with the UA of each additional component/assembly (e.g., wall, roof, etc.) determined the same way in order to calculate the overall building UA to be utilized as the base case for trade-off analysis purposes.

N1102.1 and N1102.1.3

Insulation and Fenestration Criteria (N1102.1) and Total UA Alternative (N1102.1.3)

N1102.1 and N1102.1.3 continues

N1102.1 and N1102.1.3 continued

TABLE N1102.1 Insulation and Fenestration Requirements by Component[a]

Climate Zone	Fenestration U-Factor	Skylight[b] U-Factor	Glazed Fenestration SHGC	Ceiling R-Value	Wood Frame Wall R-Value	Mass Wall R-Value	Floor R-Value	Basement[c] Wall R-Value	Slab[d] R-Value & Depth	Crawl Space[c] Wall R-Value
1	1.20	0.75	0.40	30	13	3	13	0	0	0
2	0.75	0.75	0.40	30	13	4	13	0	0	0
3	0.65	0.65	0.40[e]	30	13	5	19	0	0	5/13
4 except Marine	0.40	0.60	NR	38	13	5	19	10/13	10,2 ft	10/13
5 and Marine 4	0.35	0.60	NR	38	19 or 13 + 5[g]	13	30[f]	10/13	10,2 ft	10/13
6	0.35	0.60	NR	49	19 or 13 + 5[g]	15	30[f]	10/13	10,4 ft	10/13
7 and 8	0.35	0.60	NR	49	21	19	30[f]	10/13	15,4 ft	10/13

a. R-values are minimums. U-factors and SHGC are maximums. R-19 insulation shall be permitted to be compressed into a 2 × 6 cavity.

b. The fenestration U-factor column excludes skylights. The SHGC column applies to all glazed fenestration.

c. The first R-value applies to continuous insulation, the second to framing cavity insulation; either insulation meets the requirement.

d. R-5 shall be added to the required slab edge R-values for heated slabs.

e. There are no SHGC requirements in the Marine zone.

f. Or insulation sufficient to fill the framing cavity. R-19 minimum.

g. "13 + 5" means R-13 cavity insulation plus R-5 insulated sheathing. If structural sheathing covers 25% or less of the exterior, R-5 sheathing is not required where structural sheathing is used. If structural sheathing covers more than 25% of exterior, structural sheathing shall be supplemented with insulated sheathing of at least R-2.

CHANGE TYPE. Modification

CHANGE SUMMARY. New alternative U-factors and R-values for mass walls have been introduced.

2006 CODE: N1102.1.2 U-Factor Alternative. An assembly with a U-factor equal to or less than that specified in Table N1102.1.2 shall be permitted as an alternative to the R-value in Table N1102.1.

Exception: For mass walls not meeting the criterion for insulation location in Section N1102.2.3, the U-factor shall be permitted to be:

a. U-factor of 0.17 in Climate Zone 1

b. U-factor of 0.14 in Climate Zone 2

c. U-factor of 0.12 in Climate Zone 3

d. U-factor of 0.10 in Climate Zone 4 except Marine

e. U-factor of 0.082 in Climate Zone 5 and Marine 4

N1102.2.3 Mass Walls. Mass walls, for the purposes of this chapter, shall be considered walls of concrete block, concrete, insulated concrete form (ICF), masonry cavity, brick (other than brick veneer), earth (adobe, compressed earth block, rammed earth), and solid timber/logs. The provisions of Section N1102.1 for mass walls shall be applicable when at least 50% of the required insulation R-value is on the exterior of, or integral to, the wall. Walls that do not meet this cri-

N1102.1.2 and N1102.2.3 continues

N1102.1.2 and N1102.2.3

U-Factor Alternative (N1102.1.2) and Mass Walls (N1102.2.3)

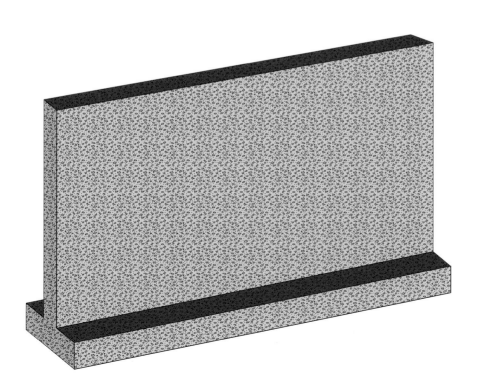

N1102.1.2 and N1102.2.3 continued terion for insulation placement shall meet the wood frame wall insulation requirements of Section N1102.1.

> **Exception:** For walls that do not meet this criterion for insulation placement the minimum added insulation *R*-value shall be permitted to be:
>
> **a.** *R*-value of 4 in Climate Zone 1
> **b.** *R*-value of 6 in Climate Zone 2
> **c.** *R*-value of 8 in Climate Zone 3
> **d.** R-value of 10 in Climate Zone 4 except Marine
> **e.** R-value of 13 in climate Zone 5 and Marine 4

CHANGE SIGNIFICANCE. The U-factor alternative method and the R-values for mass walls in the 2006 code were developed such that there would be minimal changes in the stringency of the code application for envelope provisions, in comparison with the 2003 code. Additional modifications were approved during the process, including clarification of the equivalent U-factors for mass walls with insulation on the interior of the wall, versus the alternative U-factors for wood frame walls in Section N1102.1.2. The coordinated change to Section N1102.1.2 for values in "Climate Zone 4 except Marine" and "Climate Zones 5 and Marine 4," however, was not originally included. As such, additional changes for these two climate zones were proposed and approved during the 2004/2005 code change cycle to include the alternative U-factors for mass walls with insulation on the interior in "Climate Zone 4 except Marine" and "Climate Zones 5 and Marine 4," consistent with the values for wood-frame walls.

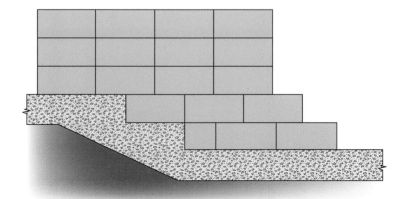

PART 5

Mechanical

Chapters 12 Through 23

■ **Chapter 12** Mechanical Administration No changes addressed

■ **Chapter 13** General Mechanical System Requirements

■ **Chapter 14** Heating and Cooling Equipment

■ **Chapter 15** Exhaust Systems

■ **Chapter 16** Duct Systems

■ **Chapter 17** Combustion Air No changes addressed

■ **Chapter 18** Chimneys and Vents No changes addressed

■ **Chapter 19** Special Fuel-Burning Equipment No changes addressed

■ **Chapter 20** Boilers and Water Heaters No changes addressed

■ **Chapter 21** Hydronic Piping

■ **Chapter 22** Special Piping and Storage Systems No changes addressed

■ **Chapter 23** Solar Systems No changes addressed

The IRC, as a code specific to the entire construction of residential buildings that fall under its scope, contains provisions for mechanical, fuel gas, plumbing, and electrical systems of the building. These systems are covered in various parts of the IRC, beginning with Part 5.

This part contains administrative provisions unique to the application and enforcement of mechanical systems, as well as the technical provisions related to the design and installation of mechanical systems. Chapter 13 provides the general requirements for all mechanical systems and addresses the labeling of appliances, types of fuel used, access to appliances for repair and maintenance, and other issues such as clearances from combustibles. The remainder of Part 5 deals with issues of specific mechanical systems such as heating and cooling systems, exhaust systems, ducts, boilers, and hydronic piping. Part 5 also contains two chapters specific to oil tanks, oil pumps, and solar systems. ■

M1305.1

Appliance Access for Inspection, Service, Repair, and Replacement

M1305.1.3 AND M1305.1.4

Appliances in Attics (M1305.1.3) and Appliances under Floor (M1305.1.4)

M1308.3

Foundations and Supports

M1411.3.1 AND M1411.3.1.1

Auxiliary and Secondary Drain System (M1411.3.1) and Water Level Monitoring Devices (M1411.3.1.1)

M1411.4

Auxiliary Drain Pan

M1501.1 AND M1506.2

Outdoor Discharge (M1501.1) and Recirculation of Air (M1506.2)

M1502.6

Duct Length

M1601.3.1

Joints and Seams

TABLES M2101.1 AND M2101.9, AND SECTIONS M2103.1, M2103.2, AND M2104.2

Hydronic Piping Materials (Table M2101.1), Hanger Spacing Intervals (Table M2101.9), Piping Materials (M2103.1), Piping Joints (M2103.2), and Piping Joints for Low-Temperature Piping (M2104.2)

M1305.1

Appliance Access for Inspection, Service, Repair, and Replacement

CHANGE TYPE.   Clarification

CHANGE SUMMARY.   The change clarifies the dimensions and the level-surface requirement of the working space for appliances. Criteria for ready access are expanded to include elements that should not be obstructing the access to the appliance for service and repair.

2006 CODE: M1305.1 Appliance Access for Inspection, Service, Repair, and Replacement.   Appliances shall be accessible for inspection, service, repair, and replacement without removing permanent construction, <u>other appliances, or any other piping or ducts not connected to the appliance being inspected, serviced, repaired, or replaced</u>. ~~Thirty inches of~~ <u>A level</u> working space <u>at least 30 inches (762 mm) deep and 30 inches (762 mm) wide</u> shall be provided in front of the control side to service an appliance. Installation of room heaters shall be permitted with at least an 18-inch (457 mm) working space. A platform shall not be required for room heaters.

CHANGE SIGNIFICANCE.   The 2006 code clarifies the language regarding ready access to appliances for inspection, service, or replacement by including other appliances, piping, and ducts in the list of items that should not be obstructing such access. Even though other appliances and other elements such as piping or duct work that are not part of the appliance being serviced may not always be considered as "permanent construction," they should not be obstructing the access to appliances. Many times, a simple and short inspection or

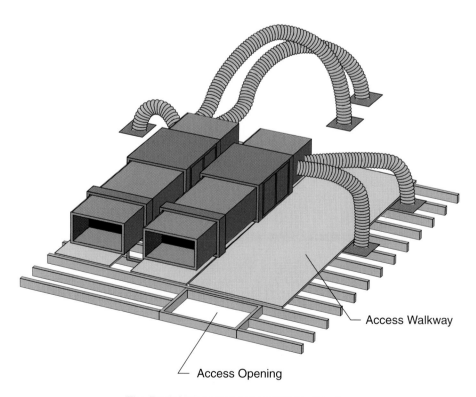

The Back Unit Is Not Accessible for Service

service call turns into a long and costly service call because features unrelated to the appliance being inspected or serviced such as piping and ductwork are in the way and must be removed or relocated so that access becomes available for the service or removal of the appliance. The new language also clarifies that the working space required for access to appliances is a level area 30 inches deep by 30 inches wide. It is obvious that surfaces that are not level and spaces that might be 30 inches in one dimension but much smaller in another could create a major problem for service and inspection personnel. The 2006 code's new language will result in construction and installation practices that ultimately intend to save time and keep costs lower for the inspection and repair of appliances.

M1305.1.3 and M1305.1.4

Appliances in Attics (M1305.1.3) and Appliances under Floor (M1305.1.4)

CHANGE TYPE. Addition

CHANGE SUMMARY. A new exception has been added that allows the access path to mechanical equipment in attics and under floor spaces to be longer than 20 feet. It allows up to 50 feet for attics and unlimited length for under-floor spaces.

2006 CODE: M1305.1.3 Appliances in Attics. Attics containing appliances requiring access shall be provided with an opening and a clear and unobstructed passageway large enough to allow removal of the largest appliance, but not less than 30 inches (762 mm) high and 22 inches (559 mm) wide and not more than 20 feet (6096 mm) in length when measured along the centerline of the passageway from the opening to the appliance. The passageway shall have continuous solid flooring in accordance with Chapter 5 not less than 24 inches (61 mm) wide. A level service space at least 30 inches (762 mm) deep and 30 inches (762 mm) wide shall be present along all sides of the appliance where access is required. The clear access opening dimensions shall be a minimum of 20 inches by 30 inches (508 mm by 762 mm), where such dimensions are large enough to allow removal of the largest appliance.

Exceptions:
1. The passageway and level service space are not required where the appliance is capable of being serviced and removed through the required opening.

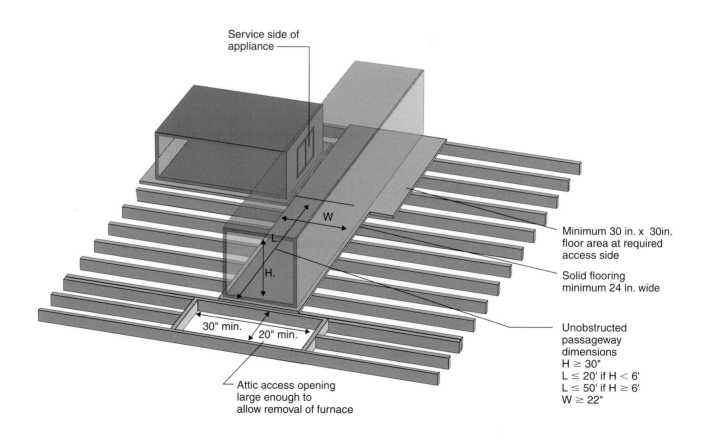

Service side of appliance

W

L

H.

30" min.   20" min.

Attic access opening large enough to allow removal of furnace

Minimum 30 in. x 30in. floor area at required access side

Solid flooring minimum 24 in. wide

Unobstructed passageway dimensions
H ≥ 30"
L ≤ 20' if H < 6'
L ≤ 50' if H ≥ 6'
W ≥ 22"

2. Where the passageway is unobstructed and not less than 6 feet (1829 mm) high and 22 inches (559 mm) wide for its entire length, the passageway shall be not greater than 50 feet (15250 mm) long.

M1305.1.4 Appliances under Floors.   Under floor spaces containing appliances requiring access shall be provided with an unobstructed passageway large enough to remove the largest appliance, but not less than 30 inches (762 mm) high and 22 inches (559 mm) wide, nor more than 20 feet (6096 mm) in length when measured along the centerline of the passageway from the opening to the appliance. A level service space at least 30 inches (762 mm) deep and 30 inches (762 mm) wide shall be present at the front or service side of the appliance. If the depth of the passageway or the service space exceeds 12 inches (305 mm) below the adjoining grade, the walls of the passageway shall be lined with concrete or masonry extending 4 inches (102 mm) above the adjoining grade in accordance with Chapter 4. The rough framed access opening dimensions shall be a minimum of 22 inches by 30 inches (559 mm by 762 mm), where the dimensions are large enough to remove the largest appliance.

Exceptions:
1. The passageway is not required where the level service space is present when the access is open, and the appliance is capable of being serviced and removed through the required opening.

2. Where the passageway is unobstructed and not less than 6 feet high (1929 mm) and 22 inches (559 mm) wide for its entire length, the passageway shall not be limited in length.

CHANGE SIGNIFICANCE.   The addition of the new exception 2 is made to correlate and make consistent the attic and under-floor equipment access provisions of the IRC with those found in the International Fuel Gas Code (IFGC). There is no reason that access requirements for electrical and gas mechanical units should be different. The 6-foot-high, 22-inch-wide unobstructed access to equipment is intended to allow inspection and service personnel to walk erect for safer and unimpeded travel, and therefore the 20-foot length limitation for such accessways is extended to 50 feet for attics and is unlimited in under-floor spaces.

M1308.3

Foundations and Supports

CHANGE TYPE. Addition

CHANGE SUMMARY. A new requirement for raising the outdoor mechanical systems above the finished grade has been introduced. The criteria introduced are similar to the provisions for heat pumps found in Section M1403.2.

2006 CODE: **M1308.3 Foundations and Supports.** Foundations and supports for outdoor mechanical systems shall be raised at least 3 inches (76 mm) above the finished grade and shall also conform to the manufacturer's installation instructions.

CHANGE SIGNIFICANCE. All mechanical units installed outdoors are subject to corrosion and other environmental effects. The installation of such units on the grade could subject them to additional and unnecessary effects of drainage water and possible ponded water. Heat pumps are essentially refrigeration units and are defined in the IRC as appliances having heating or heating/cooling capability and that use refrigerants to extract heat from air, liquid, or other sources. Accordingly, the 2003 code requires the outdoor units of heat pumps to be raised 3 inches above the finished grade because heat pump units produce defrost water, which should be allowed to drain around and under the unit and away from it. Even though mechanical units, other than heat pumps, may not produce defrost water, water from other sources such as storm surface drainage could cause the same problems for units installed on grade. As such, the 2006 code has now added similar language to raise all outdoor mechanical units 3 inches above finished grade to protect all outdoor units and not just heat pumps.

CHANGE TYPE.   Addition

CHANGE SUMMARY.   A fourth and new method of detecting blockage in the primary drain line of cooling coils or evaporators has been added. This option allows the incorporation of a water level detection device. Additionally, a new subsection M1411.3.1.1 has been introduced to address the overflow issues for downflow units.

2006 CODE: M1411.3.1 Auxiliary and Secondary Drain Systems. In addition to the requirements of Section M1411.3, a secondary drain or auxiliary drain pan shall be required for each cooling or evaporator coil where damage to any building components will occur as a result of overflow from the equipment drain pan or stoppage in the condensate drain piping. <u>Such piping shall maintain a minimum horizontal slope in the direction of discharge of not less than one-eighth unit vertical in 12 units horizontal (1-percent slope).</u> Drain piping shall be a minimum of ¾-inch (19 mm) nominal pipe size. One of the following methods shall be used:

1. An auxiliary drain pan with a separate drain shall be provided under the coils on which condensation will occur. The auxiliary pan drain shall discharge to a conspicuous point of disposal to alert occupants in the event of a stoppage of the primary drain. The pan shall have a minimum depth of 1.5 inches (38 mm), shall not be less than 3 inches (76 mm) larger than the unit or the coil dimensions in width and length, and shall be constructed of corrosion-resistant material. Metallic pans shall have a minimum thickness of not less than 0.0276-inch (0.7 mm) galvanized sheet metal. Nonmetallic pans shall have a minimum thickness of not less than 0.0625 inch (1.6 mm).

M1411.3.1 and M1411.3.1.1 continues

M1411.3.1 and M1411.3.1.1

Auxiliary and Secondary Drain System (M1411.3.1) and Water Level Monitoring Devices (M1411.3.1.1)

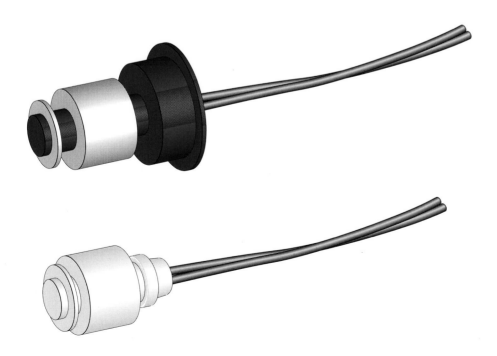

M1411.3.1 and M1411.3.1.1 continued

2. A separate overflow drain line shall be connected to the drain pan provided with the equipment. Such overflow drain shall discharge to a conspicuous point of disposal to alert occupants in the event of a stoppage of the primary drain. The overflow drain line shall connect to the drain pan at a higher level than the primary drain connection.

3. An auxiliary drain pan without a separate drain line shall be provided under the coils on which condensate will occur. This pan shall be equipped with a water level detection device conforming to UL 508 that will shut off the equipment served prior to overflow of the pan. The auxiliary drain pan shall be constructed in accordance with Item 1 of this section.

4. A water level detection device conforming to UL 508 shall be provided that will shut off the equipment served in the event that the primary drain is blocked. The device shall be installed in the primary drain line, the overflow drain line or the equipment-supplied drain pan, located at a point higher than the primary drain line connection and below the overflow rim of such pan.

M1411.3.1.1 Water Level Monitoring Devices On downflow units and all other coils that have no secondary drain and no means to install an auxiliary drain pan, a water level monitoring device shall be installed inside the primary drain pan. This device shall shut off the equipment served in the event that the primary drain becomes restricted. Externally installed devices and devices installed in the drain line shall not be permitted.

CHANGE SIGNIFICANCE. Cooling or evaporating coils produce condensate drain that must be discharged to an approved place of disposal. An auxiliary drain pan or a secondary drain is required for equipment located where condensate overflow would cause damage to a building or its contents. The purpose of the auxiliary drain pan and secondary drain is to catch condensate spilling from the primary drain pan located within the equipment. This backup protects the building from structural and finish damage. Condensate drains are notorious for clogging because of debris (lint, dust) from air-handling systems and the natural affinity to produce slime growths in drain pans and pipes. Because it is relatively common for condensate overflows to cause damage to buildings, the 2003 code lists three options for preventing damage where the equipment is located in spaces such as attics, above suspended ceilings and furred spaces, and locations on upper stories. The 2006 code has now added a fourth option. This option allows the installation of a water-level-detection device that will shut off the equipment if the primary drain is blocked. This technology is available in the market and provides the level of protection intended.

The new section M1411.3.1.1 is intended to require protection against overflow on all down flow units that do not have a secondary drain and have no provisions for installing an auxiliary drain pan. A downflow unit is a unit that blows the air downward.

M1411.4

Auxiliary Drain Pan

CHANGE TYPE. Addition

CHANGE SUMMARY. Category IV condensing appliances are now required to be provided with an auxiliary drain pan.

2006 CODE: M1411.4 Auxiliary Drain Pan. Category IV condensing appliances shall be provided with an auxiliary drain pan where damage to any building component will occur as a result of stoppage in the condensate drain piping system. These pans shall be installed in accordance with the applicable provisions of Sections M1411.3.

> **Exception:** Fuel-fired appliances that automatically shut down operation in the event of stoppage in the condensate drain system.

M1411.4 continues

Secondary Drain Pan

Provide Drain Pipe or a Water Level Detection Device

M1411.4 continued

CHANGE SIGNIFICANCE. Evaporators and cooling coils are required to be provided with auxiliary drain pans, in accordance with Section M1411.3.1, but these are no longer the only types of equipment that produce condensate. All condensing furnaces are equipped with exhaust fans with drain ports that require drains to be discharged. Most manufacturers of condensing furnaces recommend a pan under their equipment because furnace drains are just as subject to stoppage and obstructions as cooling or evaporator coils. The 2006 new text brings the code in line with the recommendations of most manufacturers in a clear and enforceable language. This requirement is applicable only where damage to building components is likely, such as installations in attics or furred spaces.

CHANGE TYPE. Clarification

CHANGE SUMMARY. New section requires air removed by exhaust fans to be discharged directly to the outside and in a location where it could not be drawn back into the house. Attics, soffits, ridge vents, and crawl spaces are not allowed as locations to discharge air from exhaust.

2006 CODE: 1501.1 Outdoor Discharge. The air removed by every mechanical exhaust system shall be discharged to the outdoors. Air shall not be exhausted into an attic, soffit, ridge vent or crawl space.

> **Exception:** Whole-house ventilation-type attic fans that discharge into the attic space of dwelling units having private attics shall be permitted.

M1506.2 Recirculation of Air. Exhaust air from bathrooms and toilet rooms shall not be recirculated within a residence or to another dwelling unit and shall be exhausted directly to the outdoors. Exhaust air from bathrooms and toilet rooms shall not discharge into an attic, crawl space, or other areas inside the building.

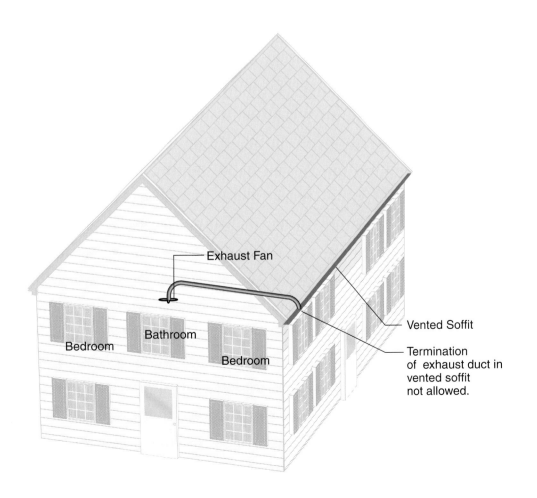

M1501.1 and M1506.2 continues

M1501.1 and M1506.2

Outdoor Discharge (M1501.1) and Recirculation of Air (M1506.2)

M1501.1 and M1506.2 continued

CHANGE SIGNIFICANCE. The new 2006 code text makes it clear that it is not intended for moisture from bath exhausts to be discharged into an attic, crawl space, or similar areas within these spaces. Whereas other sections of this chapter provide clear language in this regard for specific systems such as clothes dryers, range hoods, and overhead exhaust hoods, this chapter in the 2003 code is silent on toilet or bathroom exhaust, whole-house fans, and other environmental-exhaust systems such as from laundry rooms, spas, or saunas. As a result, it has sometimes been interpreted that discharging into soffits or ridge vents is the same as discharging outside into the atmosphere and that soffits or ridge vents or the areas in their vicinity are appropriate locations for discharge of exhaust air. To discharge most mechanical environmental exhaust systems (which frequently discharge moisture-laden air) into attics, crawl spaces, or other building cavities is not acceptable. The new language for this section of the IRC is taken from Section 501.2 of the International Mechanical Code (IMC).

CHANGE TYPE.   Modification

CHANGE SUMMARY.   Section M1502 and its subsections have been reorganized for an easier-to-follow format. The maximum length of a clothes dryer exhaust duct is limited to 25 feet, with some exceptions. Exception 1 of the 2003 code, which allows booster fans to be installed for longer lengths, has been deleted, and a new exception based on the American Society of Heating, Refrigerating, and Air-Conditioning Engineers (ASHRAE) handbook's engineered calculations has been added.

2006 CODE: M1502.6 Duct Length.   The maximum length of a clothes dryer exhaust duct shall not exceed 25 feet (7620 mm) from the dryer location to the wall or roof termination. The maximum length of the duct shall be reduced 2.5 feet (762 mm) for each 45-degree (0.79 rad) bend and 5 feet (1524 mm) for each 90-degree (1.6 rad) bend. The maximum length of the exhaust duct does not include the transition duct.

Exceptions:
1. ~~Where a clothes dryer booster fan is installed and listed and labeled for the application, the maximum length of the exhaust duct, including any transition duct, shall be permitted to be in accordance with the booster fan manufacturer's in-~~

M1502.6 continues

M1502.6
Duct Length

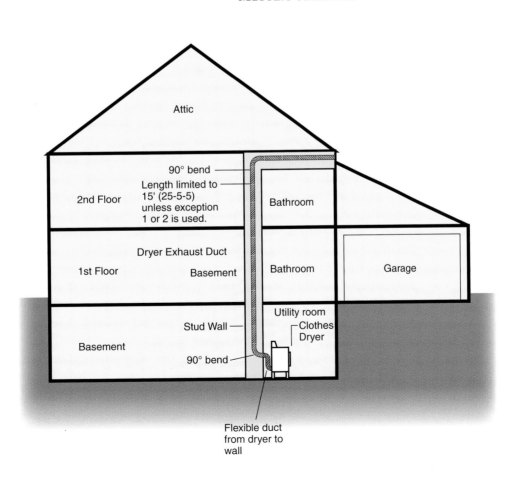

M1502.6 continued

~~stallation instructions. Where a clothes dryer booster fan is installed and not readily accessible from the room in which the dryer is located, a permanent identifying label shall be placed adjacent to where the exhaust duct enters the wall. The label shall bear the words, This dryer exhaust system is equipped with a remotely located booster fan.~~

~~2.~~ **1.** Where the make and model of the clothes dryer to be installed is known and the manufacturer's installation instructions for such dryer are provided to the building official, the maximum length of the exhaust duct, including any transition duct, shall be permitted to be in accordance with the dryer manufacturer's installation instructions.

2. Where large radius 45-degree (0.8 rad) and 90-degree (1.6 rad) bends are installed, determination of the equivalent length of clothes dryer exhaust duct for each bend by engineering calculation in accordance with ASHRAE Fundamentals Handbook shall be permitted.

CHANGE SIGNIFICANCE. Section M1501 in the 2003 code consists of three subsections. Subsection M1501.1 contains a number of requirements that would be easier to follow if they were separated under their own titles. As such, the 2006 code contains a format change with no technical implications that creates six subsections for easier use of the code, and the section number has changed to M1502. The substantive and technical changes occur in the 2003 code Section M1501.3, which is now numbered Section M1502.6 and retitled "Duct length" instead of "Length limitation."

This technical change is based on the observation that most homeowners do not routinely clean their clothes dryer exhaust ducts; therefore, there is potential for fires to occur related to improper maintenance of such systems. The 2003 code exception 1 allows use of a booster fan that creates additional maintenance needs, to extend the length of clothes dryer exhaust ducts. This could increase the potential for fires, when not properly maintained. Additionally, it is not possible to meet the requirement that booster fans be "listed and labeled for the application," because a listing for this specific application is not available. There is also a problem with the use of booster fans with most dryers, because clothes dryers themselves are required to be listed and labeled and most do not include instructions for incorporation of booster fans within the exhaust system. The remaining exception, which is based on the manufacturers' allowance of longer duct lengths, works for most any building layout because dryers from major manufacturers allow up to twice the length allowed in the charging paragraph of this section of the code. For these reasons, the 2006 code has deleted the exception for the use of booster fans.

CHANGE TYPE. Modification

CHANGE SUMMARY. Mechanical fasteners for use with flexible nonmetallic air ducts serving heating, cooling, and ventilation equipment are now required to comply with UL181B and must be marked as such.

2006 CODE: M1601.3.1 Joints and Seams. Joints of duct systems shall be made substantially airtight by means of tapes, mastics, gasketing or other approved closure systems. Closure systems used with rigid fibrous glass ducts shall comply with UL 181A and shall be marked 181A-P for pressure-sensitive tape, 181A-M for mastic or 181A-H for heat-sensitive tape. Closure systems used with flexible air ducts and flexible air connectors shall comply with UL 181B and shall be marked 181B-FX for pressure-sensitive tape or 181B-M for mastic. Duct connections to flanges of air distribution system equipment or sheet metal fittings shall be mechanically fastened. <u>Mechanical fasteners for use with flexible nonmetallic air ducts shall comply with UL181B and shall be marked 181B-C.</u> Crimp joints for round ducts shall have a contact lap of at least 1.5 inches (38 mm) and shall be mechanically fastened by means of at least three sheet metal screws or rivets equally spaced around the joint.

M1601.3.1
Joints and Seams

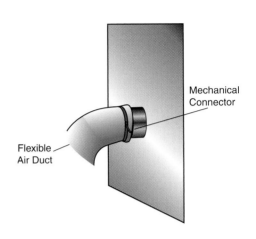

Flexible Air Duct — Mechanical Connector

CHANGE SIGNIFICANCE. The 2003 code has a void with respect to mechanical fasteners used for connecting flexible nonmetallic air ducts, because at the time of publication of the 2003 IRC, UL Standard 181 did not contain such provisions. In May 2003 new requirements were adopted in UL 181B for evaluating mechanical fasteners for such ducts, designated as UL 181B Part III. This change in the 2006 code recognizes this addition to the standard and fills the previous void by requiring these fasteners to comply with UL 181B and be marked to show compliance.

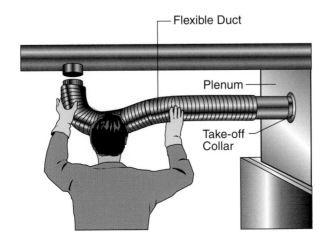

Flexible Duct — Plenum — Take-off Collar

Tables M2101.1 and M2101.9, and Sections M2103.1, M2103.2, and M2104.2

Hydronic Piping Materials (Table M2101.1), Hanger Spacing Intervals (Table M2101.9), Piping Materials (M2103.1), Piping Joints (M2103.2), and Piping Joints for Low-Temperature Piping (M2104.2)

CHANGE TYPE.   Addition

CHANGE SUMMARY. New provisions and standards for cross-linked polyethylene (PEX) tubing and joints and new provisions and standards for polypropylene pipe and tubing for use in hydronic systems have been added.

2006 CODE:

M2103.1 Piping Materials.   Piping for embedment in concrete or gypsum materials shall be standard-weight steel pipe, copper tubing, cross-linked polyethylene/aluminum/cross-linked polyethylene (PEX-AL-PEX) pressure pipe chlorinated polyvinyl chloride (CPVC), ~~or~~ polybutylene, <u>cross-linked polyethylene (PEX) tubing, or polypropylene (PP)</u> with a minimum rating of 100 psi at 180° F (690 kpa at 82° C).

M2103.2 Piping Joints.   Piping joints that are embedded shall be installed in accordance with the following requirements.

1. Steel pipe joints shall be welded.
2. Copper tubing shall be joined with brazing material having a melting point exceeding 1000° F (538° C).
3. Polybutylene pipe and tubing joints shall be installed with socket-type heat-fused polybutylene fittings.
4. CPVC tubing shall be joined using solvent cement joints.

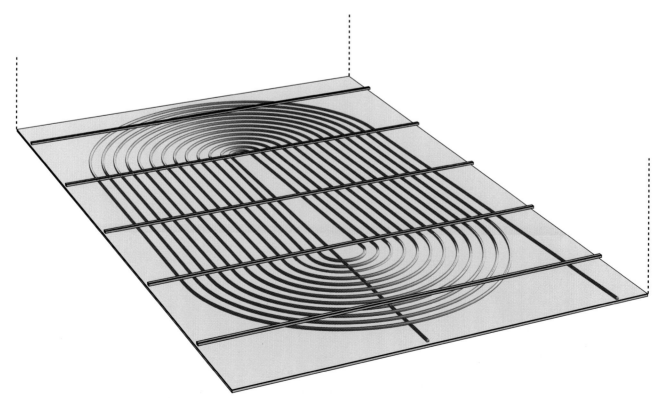

5. Polypropylene pipe and tubing joints shall be installed with socket-type heat-fusion polypropylene fittings.

6. Cross-linked polyethylene (PEX) tubing shall be joined using cold expansion, insert, or compression fittings.

TABLE M2101.1   **Hydronic Piping Materials**

Material	Use Code[a]	Standard[b]	Joints	Notes
Cross-linked Polyethylene (PEX)	1, 2, 3	ASTM F 876, F 877	~~Mechanical Compression~~ (See PEX Fittings)	Install in accordance with manufacturer's instructions
PEX Fittings		ASTM F 1807 ASTM F 1960 ASTM F 2098	Copper-Crimp/ Insert Fittings Cold Expansion Fittings Stainless Steel Clamp Insert Fittings	Install in accordance with manufacturer's instructions
Polypropylene (PP)	1, 2, 3	ISO 15874 ASTM F2389	Heat fusion joints, mechanical fittings, threaded adapters, compression joints	

(portions of table not shown remain the same)

a. Use code:
   1. Above ground.
   2. Embedded in radiant systems.
   3. Temperatures below 180° F only.
   4. Low temperature (below 130° F) only.
b. Standards as listed in Chapter 43

TABLE M2101.9   **Hanger Spacing Intervals**

(Addition to existing table shown below)

Piping Material	Maximum Horizontal Spacing (feet)	Maximum Vertical Spacing (feet)
PP < 1 inch pipe or tubing	2.67	4
PP > 1¼ inch	4	5

(portions of table not shown remain the same)

M2104.2 Piping Joints.   Piping joints (other than those in Section M2103.2) that are embedded shall be installed in accordance with the following requirements:

1. Cross-linked Polyethylene (PEX) tubing shall follow manufacturer's instructions.

2. Polyethylene tubing shall be installed with heat fusion joints.

3. Polypropylene (PP) tubing shall be installed in accordance with the manufacturer's instructions.

CHANGE SIGNIFICANCE.   The changes are related to the allowance and documenting the referenced standards related to the use of cross-linked polyethylene (PEX) in hydronic systems, because several ICC

Tables M2101.1 and M2101.9 and Sections M2103.1, M2103.2,
and M2104.2 continues

Tables M2101.1 and M2101.9 and Sections M2103.1, M2103.2, and M2104.2 continued

Evaluation reports provide for such allowances under the 1998 IMC (ER-5143 and ER-5582). Additionally, several changes have the overall purpose of allowing the use of polypropylene in hot and cold water distribution piping and radiant heating systems. Polypropylene materials meeting the indicated requirements and standards have demonstrated successful use in hot and cold water piping, radiant heating systems, chemical process piping, swimming pool circulation, compressed air systems, and irrigation. In addition to being a versatile material, polypropylene is environmentally friendly in terms of initial manufacturing, raw material usage, and energy consumption. Polypropylene can be recycled, meets the health effects criteria of National Sanitation Foundation (NSF) standard 61 without any special conditions or exemptions, and has a long history of use in these applications. Polypropylene has been used extensively in other parts of the world such as Europe and the Far East countries for over 30 years and has also been used in the United States for more demanding chemical-process piping systems.

PART 6

Fuel Gas
Chapter 24

■ **Chapter 24** Fuel Gas

Fuel gas as a special system is covered in Part 6, where issues such as fuel gas pipe design and installation, fuel gas piping materials, joints, and other such issues are addressed. The fuel gas provisions of the IRC and the text of its various sections are taken directly from the International Fuel Gas Code (IFGC) and reprinted directly into the IRC. In order to make the correlation and the coordination of the two codes easier, after each fuel gas section of the IRC the original section of the IFGC is shown in parentheses.

The fuel gas portion of the IRC contains its own specific definitions in Section G2403 in addition to the general definitions found in Chapter 2 of the IRC. Provisions, tables, and figures in other sections address the technical issues of fuel gas systems, such as appliance gas input, capacity of fuel gas pipes for system design, piping support, flow controls, gas vent systems, compressed-natural-gas dispensing systems, and other fuel gas issues. ■

G2403
Definitions: Point of Delivery

G2404.3
Listed and Labeled

G2404.10
Auxiliary Drain Pan

G2415.1
Prohibited Locations for Gas Piping

G2415.5 AND G2426.7
Protection against Physical Damage

G2415.6
Piping in Solid Floors

G2420.1.1 AND TABLE G2420.1.1
Valve Approval (G2420.1.1) and Manual Gas Valve Standards (Table G2420.1.1)

G2421.3, G2421.3.1, AND G2403
Venting of Regulators (G2421.3), Vent Piping (G2421.3.1), and Vent Piping: Breather and Relief (G2403)

G2422.1
Connecting Appliances

G2403

Definitions: Point of Delivery

CHANGE TYPE. Clarification

CHANGE SUMMARY. The point of delivery for undiluted liquefied petroleum gas systems has been clarified to be at the regulator that drops the delivery pressure down to 2 psig.

2006 CODE: Point of Delivery. For natural gas systems, the point of delivery is the outlet of the service meter assembly, or the outlet of the service regulator or service shutoff valve where a meter is not provided. Where a valve is provided at the outlet of the service meter assembly, such valve shall be considered to be downstream of the point of delivery. For undiluted liquefied petroleum gas systems, the point of delivery shall be considered to be the outlet of the first~~-stage pressure~~ regulator that ~~provides utilization pressure, exclusive of line gas regulators, in the system.~~ reduces pressure to 2 psig (13.8 kPag) or less.

CHANGE SIGNIFICANCE. The 2003 code definition has created some confusion by implying that the first-stage regulator off the liquefied petroleum (LP) tank, which typically drops the pressure to 10 psig, is the point of delivery. This is incorrect, because the point of delivery should be the outlet of what is usually referred to as the second-stage regulator. LP systems can be set up to deliver gas to customers at reduced pressures of 14 inches of water column or less and 2 psig, in which case the outlet of the "2 psig service regulator" is the point of delivery. For clarification and elimination of this inconsistent application, the 2006 code has revised the language to fix the point of delivery at the outlet of the first regulator that reduces the pressure to 2 psig or less.

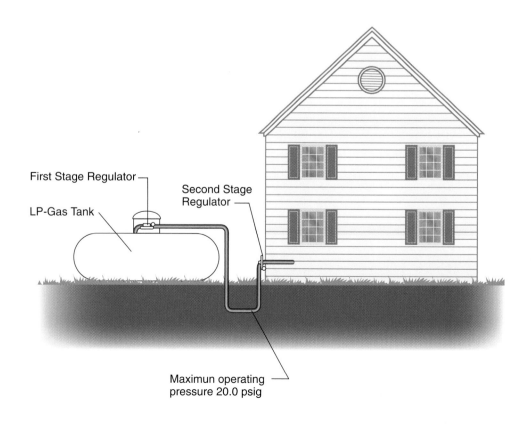

First Stage Regulator

LP-Gas Tank

Second Stage Regulator

Maximun operating pressure 20.0 psig

CHANGE TYPE. Clarification

CHANGE SUMMARY. A new sentence has been added to clarify that the listing of gas appliances must be for the applicable use.

2006 CODE: G2404.3 Listed and Labeled. Appliances regulated by this code shall be listed and labeled <u>for the application in which they are used</u> unless otherwise approved in accordance with Section R104.11. The approval of unlisted appliances in accordance with Section R104.11 shall be based upon approved engineering evaluation.

CHANGE SIGNIFICANCE. This section in the 2003 code appears to have been used, in some cases, to promote the use of products that have not undergone any testing for the application. Some examples are products that have not been tested for a particular application but have received a listing for some other application. So, although they appear to be listed, they are not for the correct application. Appliances that might have some type of label, such as one verifying sanitation requirements, are sometimes being misrepresented as labeled for other purposes. For example, appliances that are listed for residential indoor use are being installed in outdoor locations (G2406.3 requires listing for outdoor installation or protection from outdoor environmental factors). Additionally, it appears that some appliance manufacturers promote two categories of appliances: one that is listed and labeled and one that is not. The ones that are not listed and labeled are always less expensive and usually look just like the listed and labeled ones. From an installer and home-owner view, these units look like any other unit. The code requires that such unlisted appliances must be approved based upon approved engineering evaluation. A similar change has also been made to the mechanical section of the IRC in "Section M1302.1, Listed and Labeled."

G2404.3
Listed and Labeled

G2404.10
Auxiliary Drain Pan

CHANGE TYPE. Addition

CHANGE SUMMARY. The condensate removal created by the process of combustion is addressed, and provisions for auxiliary drain pans for such systems have been added.

2006 CODE: G2404.10 Auxiliary Drain Pan. Category IV condensing appliances shall be provided with an auxiliary drain pan where damage to any building component will occur as a result of stoppage in the condensate drainage system. Such pan shall be installed in accordance with the applicable provisions of Chapter 14.

> **Exception:** An auxiliary drain pan shall not be required for appliances that automatically shut down operation in the event of a stoppage in the condensate drainage system.

CHANGE SIGNIFICANCE. Evaporators and cooling coils are required to be provided with condensate drains and auxiliary drain pans in ac-

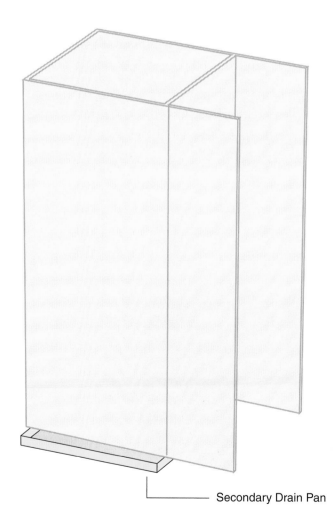

Secondary Drain Pan

Provide Drain Pipe or a Water Level Detection Device

cordance with Section M1411.3, but these are no longer the only types of appliance that produce condensate. All condensing furnaces are equipped with exhaust fans with drain ports that require drains to be discharged. Most manufacturers of condensing furnaces recommend a pan under their equipment because furnace drains are just as subject to stoppage and obstructions as cooling or evaporator coils. The 2006 new text brings the code in line with the recommendations of most manufacturers in a clear and enforceable language. This requirement is applicable only where damage to building components is likely, such as installations in attics or furred spaces. This change has also occurred in the mechanical section of the IRC, in Section M1411.4.

G2415.1

Prohibited Locations for Gas Piping

CHANGE TYPE.   Addition

CHANGE SUMMARY.   Additional criteria have been added to the "prohibited locations" section to prohibit the installation of townhouse gas piping through adjacent townhouse units.

2006 CODE: G2415.1 Prohibited Locations.   Piping shall not be installed in or through a circulating air duct, clothes chute, chimney or gas vent, ventilating duct, dumbwaiter or elevator shaft. <u>Piping installed downstream of the point of delivery shall not extend through any townhouse unit other than the unit served by such piping.</u>

CHANGE SIGNIFICANCE.   "Banking" meters at one location is very common in building construction containing multiple units. This is an efficient method of utility metering and was never envisioned to create an undesirable situation within multitenant buildings. Although gas piping has historically been installed only within the structure it serves, for townhouse construction, centralized metering location has resulted in such piping being installed within and running through multiple units of a townhouse. Having any portion of a gas piping system located in an area not accessible or controlled by the parties whom the pipe serves is not a desirable situation. In many townhouses, several lines leave the meter location and pass through the crawl spaces or the attics of the other units, leaving those portions of the system subject to damage or tampering by those in control of the other units. Additionally, townhouse units are required to be separated by two one-hour fire resistance rated wall assemblies or one two-hour fire resistance rated common wall, in accordance with Section R317.2. Each individual townhouse unit is required to be structurally independent, in accordance with Section R317.2.4. Accordingly, there has been a concern that running such pipes across multiple townhouse units not only causes multiple through penetration of the fire resistance wall assemblies, creating the need for more care and maintenance, it might also be a source of fire and explosion due to breakage of the gas piping in the event of structural failure of some part of the separation wall at a townhouse unit.

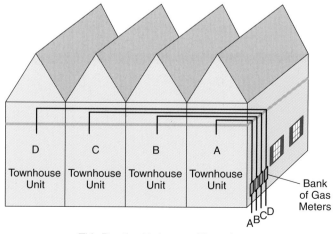

This Practice No Longer Allowed

CHANGE TYPE.   Modification and Addition

CHANGE SUMMARY.   A new section has been added to outline the protection from physical damage for gas vent pipes, and the 2003 code provisions for protection of gas piping from driven nails has been increased from 1 inch to 1.5 inches.

2006 CODE: G2415.5 Protection against Physical Damage.   In concealed locations, where piping other than black or galvanized steel is installed through holes or notches in wood studs, joists, rafters, or similar members less than ~~1~~ 1.5 inches (~~25~~ 38 mm) from the nearest edge of the member, the pipe shall be protected by shield plates. Shield plates shall be a minimum of $\frac{1}{16}$-inch-thick (1.6 mm) steel, shall cover the area of the pipe where the member is notched or bored, and shall extend a minimum of 4 inches (102 mm) above sole plates, below top plates and to each side of a stud, joist, or rafter.

G2426.7 Protection Against Physical Damage.   In concealed locations, where a vent is installed through holes or notches in studs, joists, rafters, or similar members less than 1.5 inches (38 mm) from the nearest edge of the member, the vent shall be protected by shield plates. Shield plates shall be a minimum of $\frac{1}{16}$-inch-thick (1.6 mm) steel, shall cover the area of the vent where the member is notched or bored, and shall extend a minimum of 4 inches (102 mm) above sole plates, below top plates and to each side of a stud, joist, or rafter.

G2415.5 and G2426.7 continues

G2415.5 and G2426.7

Protection against Physical Damage

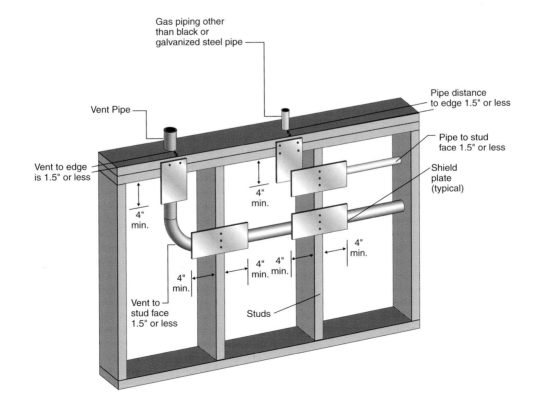

G2415.5 and G2426.7 continued

CHANGE SIGNIFICANCE. In the 2003 code, the protection of gas piping from driven nails or screws is not consistent with similar protection from the other family of International Codes. Gas piping other than black or galvanized steel located in concealed locations must be protected by steel shield plates when located closer than 1 inch to the surface. The other codes in the family of International Codes require such protection for plumbing pipes and mechanical systems when closer than 1.5 inches to the surface (see International Plumbing Code [IPC] Section 305.8, IMC Section 305.5, and IRC Sections M1308.2 and P2603.2.1) The change from 1 inch to 1.5 inches makes the protection of gas piping consistent with protection for other systems. In relation to protection from driven nails and screws for gas vent pipes, a new section, G2426.7, has been added, establishing the same criteria for gas vent piping protection as for all other systems such as gas piping and plumbing pipes. In the 2003 code, the only limit to the placing of a vent in proximity to a nailing surface is the vent's required clearance to combustible materials. When combustible materials are used, only 1-inch clearance is required for B-vents. With the use of metal framing members and/or the use of noncombustible wall covering, (i.e., cement composition tile-backer board), this limit does not apply. Additionally, with vents installed on condensing appliances, there is no required minimum clearance to combustibles. Dependence on this method for protection of vents from damage due to the installation of wall coverings, trim, decorative treatments, or pictures is not appropriate.

G2415.6
Piping in Solid Floors

CHANGE TYPE.   Clarification

CHANGE SUMMARY.   The reference to "casing" has been changed to "conduit" and a new sentence clarifies that conduits used for gas piping in solid floors must be sealed and vented to the outdoors.

2006 CODE: G2415.6 Piping in Solid Floors.   Piping in solid floors shall be laid in channels in the floor and covered in a manner that will allow access to the piping with a minimum amount of damage to the building. Where such piping is subject to exposure to excessive moisture or corrosive substances, the piping shall be protected in an approved manner. As an alternative to installation in channels, the piping shall be installed in a ~~casing~~ conduit of Schedule 40 steel, wrought iron, PVC, or ABS pipe with tightly sealed ends and joints. Both ends of such ~~casing~~ conduit shall extend not less than 2 inches (51 mm) beyond the point where the pipe emerges from the floor. <u>The conduit shall be vented above grade to the outdoors and shall be installed so as to prevent the entry of water and insects.</u>

CHANGE SIGNIFICANCE.   Gas piping installed in floors where it might be subject to excessive moisture or corrosion is required to be protected by an approved manner. One such method is the use of conduits instead of channels. The protection is provided by conduits of steel, wrought iron, polyvinyl chloride (PVC), or acrylonitrile-butadiene-styrene (ABS) pipe. The 2003 code calls these conduits "casings." Casing is not the most appropriate terminology for this purpose because it implies a conduit designed to support external loads, where typically PVC and ABS are not acceptable. Hence, the 2006 code has revised the terminology and correctly refers to it as "conduit." Additionally, a new sentence has been added to clarify that such conduits must be sealed properly, to protect the gas pipe by preventing the entry of water and insects. The conduit must also be vented to the outside in case there is leakage in the gas piping system within it.

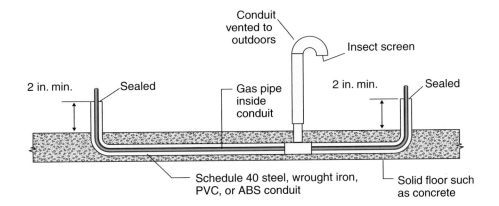

G2420.1.1 and Table G2420.1.1

Valve Approval (G2420.1.1) and Manual Gas Valve Standards (Table G2420.1.1)

CHANGE TYPE. Modification

CHANGE SUMMARY. The incomplete list of referenced standards for gas shutoff valves has been removed and a new table containing all needed standards has been inserted.

2006 CODE:

G2420.1.1 Valve Approval. Shutoff valves shall be of an approved type; shall be constructed of materials compatible with the piping; ~~Shutoff valves installed in a portion of a piping system operating above 0.5 psig or less shall comply with ASME Z 21.15 OR ASME B 16.33.~~ and shall comply with the standard that is applicable for the pressure and application, in accordance with Table G2420.1.1.

TABLE G2420.1.1 **Manual Gas Valve Standards**

		Other Valve Applications			
Valve Standards	Appliance Shutoff Valve Application Up to 1/2 psig Pressure	Up to 1/2 psig Pressure	Up to 2 psig Pressure	Up to 5 psig Pressure	Up to 125 psig Pressure
ANSI Z21.15	X				
CSA Requirement 3-88	X	X	X[a]	X[b]	
ASME B16.44	X	X	X[a]	X[b]	
ASME B16.33	X	X	X	X	X

For SI: 1 pound per square inch gauge = 6.895 KPa.

a. If labeled 2G.
b. If labeled 5G.

CHANGE SIGNIFICANCE. The standards indicated in the 2003 code text are incomplete, because they do not include any for valves used in piping systems at medium pressures (up to 5 psig). Also, valves certified under ANSI Z21.15 are for use only as appliance valves or connector shutoff valves, at or downstream of the point where the connector meets the piping system. Valves used elsewhere in the piping system must comply with one or more of the other standards. To enhance understanding and proper application of valve-certification standards, the necessary information has been presented in table form.

CHANGE TYPE. Addition and clarification

CHANGE SUMMARY. New sections have been added and existing Section G2421.3 has been revised to more clearly and accurately address the conditions under which vent pipes can and cannot be manifolded.

2006 CODE: G2421.3 Venting of Regulators. Pressure regulators that require a vent shall ~~have an independent vent to the outside of the building~~ be vented directly to the outdoors. The vent shall be designed to prevent the entry of insects, water, ~~or~~ and foreign objects.

> **Exception:** A vent to the ~~outside of the building~~ outdoors is not required for regulators equipped with and labeled for utilization with an approved vent-limiting device installed in accordance with the manufacturer's instructions.

G2421.3.1 Vent Piping. Vent piping shall be not smaller than the vent connection on the pressure regulating device. Vent piping serving relief vents and combination relief and breather vents shall be run independently to the outdoors and shall serve only a single device vent. Vent piping serving only breather vents is permitted to be connected to a manifold arrangement where sized in accordance with an approved design that minimizes back pressure in the event of diaphragm rupture.

Section G2403 General Definitions: Vent Piping. Breather. Piping run from a pressure regulating device to the outdoors, designed to provide a reference to atmospheric pressure. If the device incorporates an integral pressure relief mechanism, a breather vent can also serve as a relief vent.

    Relief. Piping run from a pressure regulating or pressure limiting device to the outdoors, designed to provide for the safe venting of gas in the event of excessive pressure in the gas piping system.

G2421.3, G2421.3.1, and G2403

Venting of Regulators (G2421.3), Vent Piping (G2421.3.1), and Vent Piping: Breather and Relief (G2403)

CHANGE SIGNIFICANCE. More clarity is needed to specify the conditions under which vent lines can and cannot be manifolded. Lines serving as relief vents must never be manifolded, because each line must be sized to carry the full release of gas from a pressure-relieving device. Some pressure regulators incorporate an integral relief mechanism, and they should also not be manifolded. Breather vents, on the other hand, serve only to provide an atmospheric reference and do not need to be sized critically for this application. Such vents can be successfully manifolded. However, in the event of a diaphragm rupture of one of the devices, manifolded vents must be sized to carry the released gas without creating excessive back pressure against the other regulating devices. Engineering methods are available to determine adequate sizing, and where such a design is presented to the code official, manifolding should be approved.

    The most common cause of the vent line obstruction is insect nests, so required protection against this problem has been explicitly added.

G2422.1

Connecting Appliances

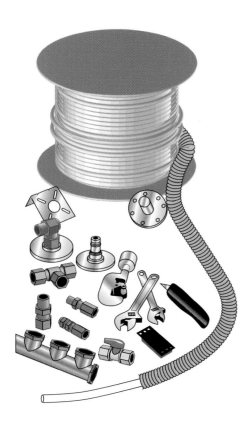

CHANGE TYPE.   Addition

CHANGE SUMMARY.   The proper standard for appliance connectors has been added, and corrugated stainless steel tubing (CSST) is now recognized as an appliance connector.

2006 CODE: G2422.1 Connecting Appliances.   Appliances shall be connected to the piping system by one of the following:

1. Rigid metallic pipe and fittings.
2. Corrugated Stainless Steel Tubing (CSST) where installed in accordance with the manufacturer's instructions.
3. Listed and labeled appliance connectors in compliance with ANSI Z21.24 and installed in accordance with the manufacturer's installation instructions and located entirely in the same room as the appliance.
4. Listed and labeled quick-disconnect devices used in conjunction with listed and labeled appliance connectors.
5. Listed and labeled gas convenience outlets used in conjunction with listed and labeled appliance connectors.
6. Listed and labeled outdoor appliance connectors in compliance with ANSI Z21.75/CSA6.27 and installed in accordance with manufacturer's installation instructions.

CHANGE SIGNIFICANCE.   The 2003 code requires that appliance connectors be listed and labeled and installed in accordance with the manufacturer's instructions, but it does not indicate what standard should be used for the listing and labeling of such appliance connectors. This void has been corrected in the 2006 code by including the proper standard for appliance connector products, ANSI Z21.24 for indoor and Z21.75/CSA6.27 for outdoor. Additionally, a new Item 2 has been added to specifically recognize CSST as a piping material that could be used to connect gas appliances to the piping system, as permitted by the CSST manufacturer's instructions.

PART 7

Plumbing
Chapters 25 Through 32

- **Chapter 25** Plumbing Administration No changes addressed
- **Chapter 26** General Plumbing Requirements No changes addressed
- **Chapter 27** Plumbing Fixtures
- **Chapter 28** Water Heaters No changes addressed
- **Chapter 29** Water Supply and Distribution
- **Chapter 30** Sanitary Drainage
- **Chapter 31** Vents
- **Chapter 32** Traps

Part 7 of the IRC contains provisions for plumbing systems. It includes a chapter on the specific and unique administrative issues of plumbing enforcement as well as the technical subjects for the overall design and installation of building plumbing systems. General plumbing issues such as the protection of piping systems from damage, piping support, and workmanship are covered in Chapter 26. The other chapters of Part 7 cover specific plumbing subjects: plumbing fixtures, water heaters, water supply and distribution, drainage, vents, and traps. ■

P2708.1 AND P2708.1.1

Shower Compartment General (P2708.1) and Access (P2708.1.1)

P2708.3

Shower Control Valves

P2713.3

Bathtub and Whirlpool Bathtub Valves

P2903.4

Thermal Expansion Control

P2904.3

Polyethylene Plastic Piping Installation

P2904.4

Water Service Pipe

P2904.5.1

Under Concrete Slabs

P2904.10

Polypropylene Plastic

TABLES P3002.1(1), P3002.1(2), P3002.2, AND P3002.3

Above-Ground Drainage and Vent Pipe (Table P3002.1(1)), Underground Building Drainage and Vent Pipe (Table P3002.1(2)), Building Sewer Pipe (Table P3002.2), and Pipe Fittings (Table P3002.3)

P3003

Joints and Connections

P3102

Vent Stacks and Stack Vents

P3103.1

Roof Extension

P3105.2

Fixture Drains

P3108.1

Wet Vent Permitted

P3108.2 AND P3108.3

Vent Connections (P3108.2) and Size (P3108.3)

P3201.6

Number of Fixtures per Trap

TABLE P3201.7

Size of Traps and Trap Arms for Plumbing Fixtures

P2708.1 and P2708.1.1

Shower Compartment General (P2708.1) and Access (P2708.1.1)

CHANGE TYPE.   Addition

CHANGE SUMMARY.   New provisions for the interior dimensions of shower compartments have been added, and the minimum width of shower access doors has been established to be 22 inches.

2006 CODE: P2708.1 General.   Shower compartments shall have at least 900 square inches (0.6 m²) of interior cross-sectional area. Shower compartments shall not be less than 30 inches (762 mm) in minimum dimension measured from the finished interior dimension of the shower compartment, exclusive of fixture valves, shower heads, soap dishes, and safety grab bars or rails. The minimum required area and dimension shall be measured from the finished interior dimension at a height equal to the top of the threshold and at a point tangent to its centerline and shall be continued to a height not less than 70 inches (1778 mm) above the shower drain outlet.

Exceptions:
1. Fold–down seats shall be permitted in the shower, provided the required 900-square-inch (0.6 m²) dimension is maintained when the seat is in the folded-up position.
2. Shower compartments having not less than 25 inches (635 mm) in minimum dimension measured from the finished interior dimension of the compartment provided the shower compartment has a minimum of 1300 square inches (0.838 m²) of cross-sectional area.

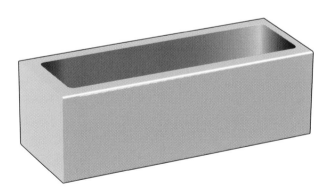

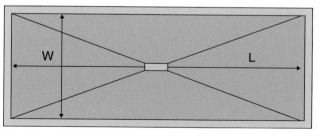

W = 25" min.
L = depends on W. It shall be a minimum of 1300 ÷ W.

Shower pan

P2708.1.1 Access. The shower compartment access and egress opening shall have a minimum clear and unobstructed finished width of 22 inches (559 mm).

CHANGE SIGNIFICANCE. The 2006 code contains a new exception 2 which allows the minimum dimension of the shower receptor to be 25 inches instead of 30 inches, as long as the overall cross-sectional area is at least 1300 square inches. This new exception introduced in code change proposal P22-04/05 was originally intended for shower compartments replacing existing tubs for the needs of the disabled and the aging population. The public comment process, however, revised the language to apply to all shower compartments to allow this flexibility for new as well as replacement units. The original proponent had included a minimum cross-sectional area of 1024 inches which was increased to 1300 square inches in the public comment process.

Additionally, the 2003 code does not provide any criteria for the minimum width of shower-compartment access doors. This has resulted in some jurisdictions being faced with approving shower-access openings as narrow as 16 ½ inches. The 2006 code has now introduced a minimum dimension of 22 inches, which is a reasonable clear width not only for functional access by the occupant but also for service and maintenance of the compartment and shower valves, as well as response and rescue should the need arise.

P2708.3

Shower Control Valves

CHANGE TYPE.   Modification

CHANGE SUMMARY.   For shower control valves, an alternative standard, CSA B125, has been introduced to what is currently referenced in the 2003 code, ASSE 1016. It is also made explicitly clear that inline thermostatic valves are not acceptable as control valves regulated by this section.

2006 CODE: P2708.3 Shower Control Valves.   Individual ~~Shower~~ shower and tub-shower combination~~s~~ valves shall be equipped with control valves of the pressure balance, the thermostatic mixing or the combination pressure balance/thermostatic mixing valve types with high limit stop in accordance with ASSE 1016 or CSA B125. The high limit stop shall be set to limit water temperature to a maximum of 120° F (49° C). Inline thermostatic valves shall not be utilized for compliance with this section.

CHANGE SIGNIFICANCE.   The Canadian Standards Association (CSA) standard CSA B125 has been added to the 2006 code for additional design considerations for shower control valves, because this standard is already acceptable and referenced in the IRC and IPC. This CSA standard has also been added to various other sections of the 2006 IRC, such as Sections P2702.2, P2722.1, and Table P2701.1.

A new sentence has now been added to eliminate the use of inline thermostatic valves as shower control valves, because these devices are unable to limit the temperature to what is required by this section. In-line thermostatic valves do not provide for thermal shock protection for individual shower applications since there is further mixing downstream, which negates the regulation of the temperature required by the referenced standards. If the cold water supply to a mixing valve is unprotected, which occurs with in-line devices installed only on the hot water supply, there will be temperature variations in the outlet of the shower head far exceeding those required by this section and its standards.

CHANGE TYPE. Addition

CHANGE SUMMARY. Temperature-limiting-device requirements have been introduced in the 2006 code, mandating that hot water supplied to bathtubs and similar fixtures be limited to 120° F.

2006 CODE: <u>**P2713.3 Bathtub and Whirlpool Bathtub Valves.**</u> <u>The hot water supplied to bathtubs and whirlpool bathtubs shall be limited to a maximum temperature of 120° F (49° C) by a water temperature limiting device that conforms to ASSE 1070, except where such protection is otherwise provided by a combination tub/shower valve in accordance with Section P2708.3.</u>

CHANGE SIGNIFICANCE. The 2003 code does not contain clear provisions to protect occupants from scalding. Hot-water scalding can cause severe burns, and the shock associated with it could cause slips and falls, thereby increasing the rate of injuries. The new language in the 2006 code provides protection from inadvertent scalding in bathtubs and in whirlpool bathtubs by mandating a water-temperature-limiting device that is listed to ASSE 1070 as a minimum code requirement and that is able to limit the hot water temperature to 120°F. The ASSE 1070 standard provides for devices to be installed appropriately with the fixture fitting or to be integral with the plumbing fixture fitting supplying the water. This is the newest ASSE standard for temperature limiting devices.

P2713.3
Bathtub and Whirlpool Bathtub Valves

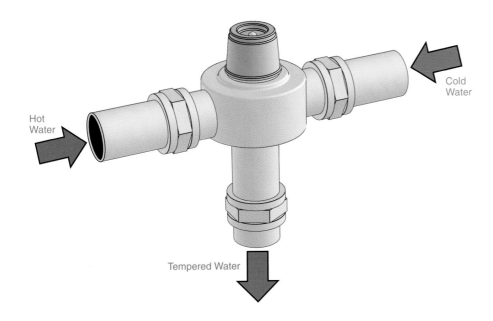

Hot Water

Cold Water

Tempered Water

P2903.4

Thermal Expansion Control

CHANGE TYPE. Modification

CHANGE SUMMARY. The provisions for thermal expansion control in water distribution systems have been completely revised by deleting the 2003 language and adding new language consistent with the IPC.

2006 CODE: ~~**P2903.4 Thermal Expansion.** In addition to the required pressure relief valve, an approved device for thermal expansion control shall be installed on any water supply system utilizing storage water heating equipment whenever the building supply pressure exceeds the pressure reducing valve setting or when any device, such as a pressure reducing valve, backflow preventer or check valve, is installed that prevents pressure relief~~ through the building supply. ~~The thermal expansion control device shall be sized in accordance with the manufacturer's installation instructions.~~

P2903.4 Thermal Expansion Control. A means for controlling increased pressure caused by thermal expansion shall be provided where required in accordance with Sections P2903.4.1 and P2903.4.2.

P2903.4.1 Pressure-Reducing Valve. For water service system sizes up to and including 2 inches (51 mm), a device for controlling pressure shall be installed where, because of thermal expansion, the pressure on the downstream side of a pressure-reducing valve exceeds the pressure-reducing valve setting.

P2903.4.2 Backflow Prevention Device or Check Valve. Where a backflow prevention device, check valve, or other device is installed on a water supply system utilizing storage water heating equipment such that thermal expansion causes an increase in pressure, a device for controlling pressure shall be installed.

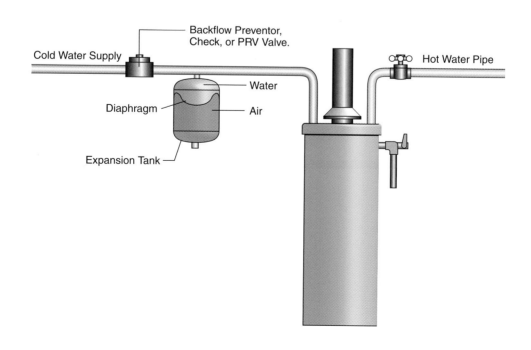

CHANGE SIGNIFICANCE.   In the connection of a water distribution system to water-heating appliances there is the potential for the migration of heated water into the water distribution piping. In a typical water distribution system, the water will expand into the water service and into the public main if the water is not withdrawn from the system at an outlet or connection. If the expansion of water is not accommodated in the system, dangerously high pressures can develop that can cause damage to piping, components, and the water heater.

The 2003 code language does not cover all possibilities in this regard and is stated in very general terms. The new language in the 2006 code is taken from IPC Section 607.3 and is intended to eliminate the confusion and provide clear guidance. It specifically addresses situations where excessive street pressures exist and provides complete and thorough provisions for thermal expansion for all potential situations. It contains two specific subsections addressing pressure-reducing valves and backflow prevention devices or check valves. Section P2903.4.1 would require either an internal bypass feature in the pressure-reducing valve or a relief valve or expansion tank installed downstream of the pressure-reducing valve. Section P2903.4.2 applies to storage-type water-heating systems because thermal expansion of water is minimal where tankless, instantaneous-type nonstorage water heaters are used. Incorporating language from the IPC makes this section of the IRC consistent with the IPC.

P2904.3

Polyethylene Plastic Piping Installation

CHANGE TYPE. Addition

CHANGE SUMMARY. Joints and joining methods for polyethylene plastic piping are now addressed in two new sections in the 2006 IRC, heat-fusion joints and mechanical joints.

2006 CODE: P2904.3 Polyethylene Plastic Piping Installation. Polyethylene pipe shall be cut square, using a cutter designed for plastic pipe. Except when joined by heat fusion pipe ends shall be chamfered to remove sharp edges. Pipe that has been kinked shall not be installed. For bends, the installed radius of pipe curvature shall be greater than 30 pipe diameters or the coil radius when bending with the coil. Coiled pipe shall not be bent beyond straight. Bends shall not be permitted within 10 pipe diameters of any fitting or valve. ~~Stiffener inserts used with compression-type fittings shall not extend beyond the clamp or nut of the fitting. Flared joints shall be permitted where recommended by the manufacturer and made by the use of a tool designed for that operation.~~ Joints between polyethylene plastic pipe and fittings shall comply with Sections P2904.3.1 and P2904.3.2.

P2904.3.1 Heat-Fusion Joints. Joint surfaces shall be clean and free from moisture. All joint surfaces shall be heated to melting temperature and joined. The joint shall be undisturbed until cool. Joints shall be made in accordance with ASTM D 2657.

P2904.3.2 Mechanical Joints. Mechanical joints shall be installed in accordance with the manufacturer's instructions.

CHANGE SIGNIFICANCE. Although there has been disagreement about the use of polyethylene plastic piping for various purposes in plumbing systems, the IRC and the IPC allow this material as specifically addressed in various sections. It is believed that one of the advantages of polyethylene plastic piping is that it can be used in "trenchless" installations in specific situations. In certain cases this might be a cost-effective method of installation, and in other cases it might help preserve existing property features, such as other underground utilities and trees. The 2003 code P2904.3 addresses polyethylene plastic piping installation but does not address the joints and joining methods for this product. Heat-fusion joints are now covered

in new Section P2904.3.1 and are required to be made in accordance with ASTM D2657. Many believe that heat fusion joints for this product are not the best because they make a slight bump on the interior of the piping system, around the entire circumference of each joint. Mechanical joints are covered in new Section P2904.3.2 and are required to comply with the manufacturer's instructions, which is consistent with mechanical joint requirements for other piping materials. Similar additions addressing polyethylene piping joints have been made in Chapter 30 of the IRC for Sanitary Drainage.

P2904.4

Water Service Pipe

CHANGE TYPE. Modification

CHANGE SUMMARY. The code now addresses conditions where water service pressure exceeds 160 psi by requiring water piping material to be rated for the highest available pressure. Also, all ductile iron water service piping are now required to be cement mortar lined.

2006 CODE: P2904.4 Water Service Pipe. Water service pipe shall conform to NSF 61 and shall conform to one of the standards listed in Table P2904.4. Water service pipe or tubing, installed underground and outside of the structure, shall have a minimum working pressure rating of 160 psi at 73° F (1103 kPa at 23° C). Where the water pressure exceeds 160 psi (1103 kPa), piping material shall have a rated working pressure equal to or greater than the highest available pressure. Water service piping materials not third-party certified for water distribution shall terminate at or before the full open valve located at the entrance to the structure. Ductile iron water service piping shall be cement mortar lined in accordance with AWWA C104.

CHANGE SIGNIFICANCE. The new language is intended to make the IRC plumbing provision for water service piping more consistent with similar IPC provisions. Further, this language is necessary because many areas have working water service pressures in excess of 160 psi, and since the 2003 code does not address pressures exceeding 160 psi, it potentially permits piping that might not be rated for the pressures available, possibly resulting in piping material failure and costly repairs. The new language allows the water service piping to remain uninterrupted until it reaches a logical place for transition, at the full open valve at the entrance to the structure. This service valve at the entrance of the water service to the structure is required by code. Additionally, all ductile iron water service pipes are now required to be cement mortar–lined by standard AWWA C104. Ductile iron mate-

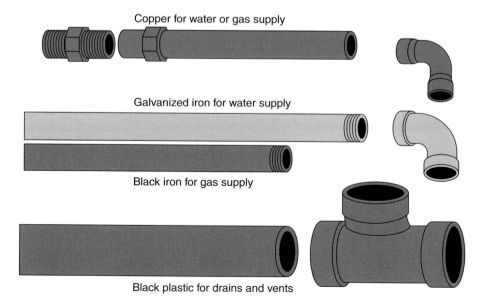

Copper for water or gas supply

Galvanized iron for water supply

Black iron for gas supply

Black plastic for drains and vents

rial standards are already covered in Table P2904.4.1 as American Water Works Association (AWWA) C115 (Standard for Flanged Ductile-iron Pipe with Ductile-iron or Gray-iron Threaded Flanges) and AWWA C151 (Standard for Ductile-iron Pipe, Centrifugally Cast, for Water). The inside surface of ductile iron water pipes are required to be cement mortar lined to provide a protective barrier between potable water in the pipe and the pipe material to prevent possible contaminants entering into the water.

P2904.5.1
Under Concrete Slabs

CHANGE TYPE.   Addition

CHANGE SUMMARY.   The code now allows polypropylene pipe or tubing under concrete slabs and for water distribution and water service systems. Polyethylene/aluminum/polyethylene (PE-AL-PE) pressure pipe, currently allowed in the 2003 code for use in water service and water distribution systems, is now specifically allowed to be used under concrete slabs.

2006 CODE: P2904.5.1 Under Concrete Slabs.   Inaccessible water distribution piping under slabs shall be copper water tube minimum Type M, brass, ductile iron pressure pipe, cross-linked polyethylene/aluminum/cross-linked polyethylene (PEX-AL-PEX) pressure pipe, <u>polyethylene/aluminum/polyethylene (PE-AL-PE) pressure pipe,</u> chlorinated polyvinyl chloride (CPVC), polybutylene (PB), cross-linked polyethylene (PEX) plastic pipe or tubing <u>or polypropylene (PP) pipe or tubing,</u> all to be installed with approved fittings or bends. The minimum pressure rating for plastic pipe or tubing installed under slabs shall be 100 psi at 180° F (689 kPa at 82° C).

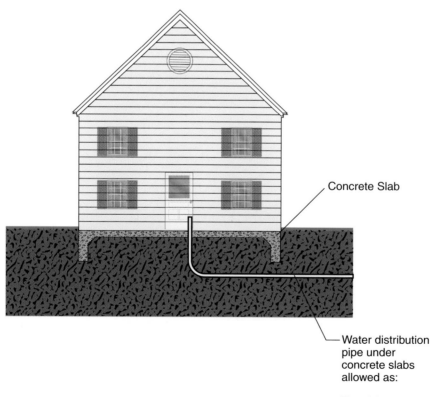

Concrete Slab

Water distribution pipe under concrete slabs allowed as:

Type M copper water tube
Brass
Dutile iron pressure pipe
PEX-AL-PEX pressure pipe
CPVC
Polybutylene
PEX pipe or tubing
PP

CHANGE SIGNIFICANCE.   The 2003 code does not allow the use of polypropylene (PP) pipe or tubing for water service or water distribution piping. As such, it is obviously not listed as a piping material that is allowed to be used under concrete slabs. Polypropylene materials meeting the proposed standards have demonstrated successful performance when used in systems such as hot and cold water piping, radiant heating systems, and chemical-process piping. As such, the 2006 code has revised Tables P2904.4.1 and P2904.5 and Section P2904.5.1 by including this piping material and now allows polypropylene pipe or tubing for uses in water service, water distribution, and under concrete slabs.

In the 2003 code, polyethylene/aluminum/polyethylene (PE-AL-PE) is allowed for use in water service and water distribution systems but not allowed for use under slabs. Polyethylene/aluminum/cross-linked polyethylene (PEX-AL-PEX) is allowed for water distribution only and is acceptable for use under concrete slabs. Accordingly, it seems appropriate that use of polyethylene/aluminum/polyethylene under concrete slabs should also be allowed because it has similar durability characteristics and has a pressure and temperature rating of 100 psi at 180° F.

P2904.10

Polypropylene Plastic

CHANGE TYPE. Addition

CHANGE SUMMARY. New language has been introduced in the 2006 code that allows the use of polypropylene plastic pipe in hot and cold water distribution systems. Two types of joints for this material, heat fusion and mechanical and compression sleeve joints, have been specifically addressed.

2006 CODE: **P2904.10 Polypropylene (PP) Plastic.** Joints between PP plastic pipe and fittings shall comply with Sections P2904.10.1 or P2904.10.2.

P2904.10.1 Heat-Fusion Joints. Heat fusion joints for polypropylene pipe and tubing joints shall be installed with socket-type heat-fused polypropylene fittings, butt-fusion polypropylene fittings or electrofusion polypropylene fittings. Joint surfaces shall be clean and free from moisture. The joint shall be undisturbed until cool. Joints shall be made in accordance with ASTM F 2389.

P2904.10.2 Mechanical and Compression Sleeve Joints. Mechanical and compression sleeve joints shall be installed in accordance with the manufacturer's installation instructions.

CHANGE SIGNIFICANCE. The new text in the 2006 code is introduced to allow the use of polypropylene in hot and cold water distribution piping and radiant heating systems. This material is required to comply with ASTM 2389, which is referenced in this section and is also incorporated into the various tables for water distribution system. The standard requires the minimum rating of 160 psi at 73° F (water service) and 100 psi at 180° F (hot and cold water distribution). Polypropylene materials meeting these requirements have had a successful history in hot and cold water piping, radiant heating systems, chemical-process piping, and other uses. Polypropylene is also an environmentally friendly material in terms of initial manufacturing, raw material usage, and energy consumption. Polypropylene can be recycled and meets the health-effects criteria of NSF 61 without any special conditions or exemptions.

CHANGE TYPE. Modification

CHANGE SUMMARY. The 2003 code Tables P3002.1 (Drain, Waste And Vent Piping And Fitting Material) and P3002.2 (Building Sewer Piping) have been deleted and replaced with four new tables, Above Ground Drainage and Vent Pipe (Table P3002.1(1)), Underground Building Drainage and Vent Pipe (Table P3002.1(2)), Building Sewer Pipe (Table P3002.2), and Pipe Fittings (Table P3002.3), for consistency with the IPC and for further ease of use.

2006 CODE: **Table P3002.1(1) Above-Ground Drainage and Vent Pipe** (see following text). **Tables P3002.1(2) Underground Building Drainage and Vent Pipe and P3002.2 Building Sewer Pipe** (see page 250). **Table P3002.3 Pipe Fittings** (see page 251).

CHANGE SIGNIFICANCE. The main change in this part of the code is to delete in its entirety the familiar Tables P3002.1 and P3002.2, addressing the materials and standards for drainage, waste, vent, and fittings, and replace them with four new tables. This reorganization makes the 2006 IRC more user-friendly and easier to follow; the 2003 code tables include a mix of information such as coupling and joining materials along with piping materials. Additionally, other technical changes proposed by several code changes have been incorporated into these tables and other related code sections such as Section P3002.2. These changes include requiring the primers for solvent-cemented PVC-DWV (drain, waste, and vent) pipe and fittings be purple and deleting cast iron pipe as an allowed piping material for forced main sewer piping. These changes will also make the IRC more consistent with the IPC.

Tables P3002.1(1), P3002.1(2), P3002.2, and P3002.3 continues

Tables P3002.1(1), P3002.1(2), P3002.2, and P3002.3

Above-Ground Drainage and Vent Pipe (Table P3002.1(1)), Underground Building Drainage and Vent Pipe (Table P3002.1(2)), Building Sewer Pipe (Table P3002.2), and Pipe Fittings (Table P3002.3)

TABLE P3002.1(1) Above-Ground Drainage and Vent Pipe

Material	Standard
Acrylonltrile butadiene styrene (ABS) plastic pipe	ASTM D 2661; ASTM F 628; CSA B181.1
Brass pipe	ASTM B 43
Cast-iron pipe	ASTM A 74; CISPI 301; ASTM A 888
Coextruded composite ABS DWV schedule 40 IPS pipe (solid)	ASTM F 1488
Coextruded composite ABS DWV schedule 40 IPS pipe (cellular core)	ASTM F 1488
Coextruded composite PVC DWV schedule 40 IPS pipe (solid)	ASTM F 1488
Coextruded composite PVC DWV schedule 40 IPS pipe (cellular core)	ASTM F 1488; ASTM F 891
Coextruded composite PVC IPS-DR, PS140, PS200 DWV	ASTM F 1488
Copper or copper-alloy pipe	ASTM B 42; ASTM B 302
Copper or copper-alloy tubing (Type K, L, M or DWV)	ASTM B 75; ASTM B 88; ASTM B 251; ASTM B 306
Galvanized steel pipe	ASTM A 53
Polyolefin pipe	CSA B181.3
Polyvinyl chloride (PVC) plastic pipe (Type DWV)	ASTM D 2665; ASTM D 2949; CSA B181.2; ASTM F 1488
Stainless steel drainage systems, Types 304 and 316L	ASME A112.3.1

Tables P3002.1(1), P3002.1(2),
P3002.2, and P3002.3 continued

TABLE R3002.1(2)   Underground Building Drainage and Vent Pipe

Material	Standard
Acrylonltrile butadiene styrene (ABS) plastic pipe	ASTM D 2661; ASTM F 628; CSA B181.1
Asbestos-cement pipe	ASTM C428
Cast-iron pipe	ASTM A 74; CISPI 301; ASTM A 888
Coextruded composite ABS DWV schedule 40 IPS pipe (solid)	ASTM F 1488
Coextruded composite ABS DWV schedule 40 IPS pipe (cellular core)	ASTM F 1488
Coextruded composite PVC DWV schedule 40 IPS pipe (solid)	ASTM F 1488
Coextruded composite PVC DWV schedule 40 IPS pipe (cellular core)	ASTM F 891; ASTM F 1488
Coextruded composite PVC IPS-DR, PS140, PS200 DWV	ASTM F 1488
Copper or copper-alloy tubing (Type K, L, M or DWV)	ASTM B 75; ASTM B 88; ASTM B 251; ASTM B 306
Polyolefin pipe	CSA B181.3; ASTM F 1412
Polyvinyl chloride (PVC) plastic pipe (Type DWV)	ASTM D 2665; ASTM D 2949; CSA B181.2
Stainless steel drainage systems, Type 316L	ASME A112.3.1

TABLE P3002.2   Building Sewer Pipe

Material	Standard
Acrylonltrile butadiene styrene (ABS) plastic pipe	ASTM D 2661; ASTM D 2751; ASTM F 628
Asbestos-cement pipe	ASTM C 428
Cast-iron pipe	ASTM A 74; ASTM A 888; CISPI 301
Coextruded composite ABS DWV schedule 40 IPS pipe (solid)	ASTM F 1488
Coextruded composite ABS DWV schedule 40 IPS pipe (cellular core)	ASTM F 1488
Coextruded composite PVC DWV schedule 40 IPS pipe (solid)	ASTM F 1488
Coextruded composite PVC DWV schedule 40 IPS pipe (cellular core)	ASTM F 1488; ASTM F 891
Coextruded composite PVC IPS-DR-PS, DWV, PS140, PS200	ASTM F 1488
Coextruded composite ABS sewer and drain DR-PS in PS35, PS50, PS100, PS140, PS200	ASTM F 1488
Coextruded composite PVC sewer and drain DR-PS in PS35, PS50, PS100, PS140, PS200	ASTM F 1488
Coextruded composite PVC sewer and drain PS 25, PS 50, PS 100 (cellular core)	ASTM F 891
Concrete pipe	ASTM C14; ASTM C 76; CSA A257.1M; CSA A267.2M
Copper or copper-alloy tubing (Type K or L)	ASTM B 75; ASTM B 88; ASTM B 251
Polyethylene (PE) plastic pipe (SDR-PR)	ASTM F 714
Polyolefin pipe	CSA B181.3; ASTM F1412
Polyvinyl chloride (PVC) plastic pipe (Type DWV, SDR26, SDR35, SDR41, PS50 or PS100)	ASTM D 2665; ASTM D 2949; ASTM D 3034; ASTM F1412; CSA B182.2; CSA B182.4
Stainless steel drainage systems, Types 304 and 316L	ASME A112.3.1
Vitrified clay pipe	ASTM C425; ASTM C700

TABLE P3002.3   Pipe Fittings

Material	Standard
Acrylonltrile butadiene styrene (ABS) plastic pipe	ASTM D 3311; CSA B181.1; ASTM D 2661
Cast iron pipe	ASME B 16.12; ASTM A 74; ASTM A 888; CISPI 301
Coextruded composite ABS DWV schedule 40 IPS pipe (solid or cellular core)	ASTM D 2661; ASTM D 3311; ASTM F 628
Coextruded composite PVC DWV schedule 40 IPS-DR, PS140, PS200 (solid or cellular core)	ASTM D 2665; ASTM D 3311; ASTM F 891
Coextruded composite ABS sewer and drain DR-PS in PS35, PS50, PS100, PS140; PS200	ASTM D 2751
Coextruded composite PVC schedule 40 IPS-DR, PS140; PS200 (solid and cellular core)	ASTM D 2665; ASTM D3311; ASTM F891
Coextruded composite PVC sewer and drain DR-PS in PS35, PS50, PS100, PS140, PS200	ASTM D3034
Copper or copper-alloy	ASME B 16.23; ASME B 16.29
Gray iron and ductile iron	AWWA C110
Polyolefin	CSA B 181.3; ASTM F1412
Polyvinyl chloride (PVC) plastic	ASTM D 3311; ASTM D 2665; ASTM F1412; ASTM F1866; CSA B 181.2; CSA B 182.4
Stainless steel drainage systems, Types 304 and 316L	ASME A 112.3.1

P3003

Joints and Connections

CHANGE TYPE.   Modification

CHANGE SUMMARY.   Subsections P3003.3 through P3003.4.4 for joints and connections have been completely deleted and substituted with Subsections P3003.3 through P3003.18. This reorganization makes the code easier to use and includes and updates the joining methods based on the type of materials for current 2003 code as well as the newly approved materials for the 2006 code.

2006 CODE: (Only a small portion of this change is shown here because of space constraints. For complete code text, refer to the 2006 IRC.)

P3003.3 ABS Plastic.   Joints between ABS plastic pipe or fittings shall comply with Sections P3003.3.1 through P3003.3.3.

P3003.3.1 Mechanical Joints.   Mechanical joints on drainage pipes shall be made with an elastomeric seal conforming to ASTM C 1173, ASTM D 3212, or CSA B602. Mechanical joints shall only be installed in underground systems unless otherwise approved. Joints shall be installed in accordance with the manufacturer's installation instructions.

P3003.3.2 Solvent Cementing.   Joint surfaces shall be clean and free from moisture. Solvent cement that conforms to ASTM D 2235 or CSA B181.1 shall be applied to all joint surfaces. The joint shall be made while the cement is wet. Joints shall be made in accordance with ASTM D 2235, ASTM D 2661, ASTM F 628, or CSA B181.1. Solvent-cement joints shall be permitted above or below ground.

P3003.3.3 Threaded Joints.   Threads shall conform to ASME B1.20.1. Schedule 80 or heavier pipe shall be permitted to be threaded with dies specifically designed for plastic pipe. Approved thread lubricant or tape shall be applied on the male threads only.

P3003.4 Asbestos-Cement.   Joints between asbestos-cement pipe or fittings shall be made with a sleeve coupling of the same composition as the pipe, sealed with an elastomeric ring conforming to ASTM D 1869.

P3003.5 Brass.   Joints between brass pipe or fittings shall comply with Sections P3003.5.1 through P3003.5.3.

P3003.5.1 Brazed Joints. All joint surfaces shall be cleaned. An approved flux shall be applied where required. The joint shall be brazed with a filler metal conforming to AWS A5.8.

P3003.5.2 Mechanical Joints. Mechanical joints shall be installed in accordance with the manufacturer's installation instructions.

P3003.5.3 Threaded Joints. Threads shall conform to ASME B1.20.1. Pipe-joint compound or tape shall be applied on the male threads only.

CHANGE SIGNIFICANCE. The 2003 code Section P3003, Joints and Connections, is not very well organized and does not contain detailed requirements or references for some piping materials. The complete re-organization of this section by deleting sections P3003.3 through P3003.4.5 and replacing them with Sections P3003.3 though P3003.18 not only makes the code easier to follow but also updates all of the re-lated materials and referenced standards for joints and connection ma-terials. There are additional technical changes incorporated into the new sections as a result of other proposed code changes. Two examples of such changes can be found in the 2006 code Section P3003.6.3, Mechanical joint coupling (ICC code change number P85-03/04), and Section P3003.16.1, Polyolefin plastic (ICC code change number P87-03/04). Some have expressed the view that the new 2006 code lan-guage is not in the format of code and contains too much detail and de-scription, such as is found in commentary. Although this might be true in some parts of the new language, it was felt that the descriptions make the application and enforcement of the code more clear and im-prove communication on this subject. These changes also have the overall effect of making the IRC plumbing provisions more consistent with the IPC.

P3102

Vent Stacks and Stack Vents

CHANGE TYPE.   Modification

CHANGE SUMMARY The requirement for stack vents has been deleted because IRC structures are limited to 3 stories and vent stacks would never be required. Additional re-formatting has taken place to eliminate vague terminology, such as "undiminished" and "directly as possible."

2006 CODE: P3102.1 Stack Required. ~~Every building shall have a vent stack or stack vent. Such vent shall run undiminished in size and as directly as possible from the building drain through to the open air above the roof.~~

P3102.2 Vent Connection to Drainage Stack. ~~Every vent stack shall connect to the base of the drainage stack. The vent stack shall connect at or below the lowest horizontal branch. Where the vent stack connects to the building drain, the connection shall be located within 10 pipe diameters downstream of the drainage stack. A stack vent shall be a vertical extension of the drainage stack.~~

P3102.3 Vent Termination. ~~Every vent stack or stack vent shall terminate outdoors to the open air or terminate to a stack-type air admittance valve.~~

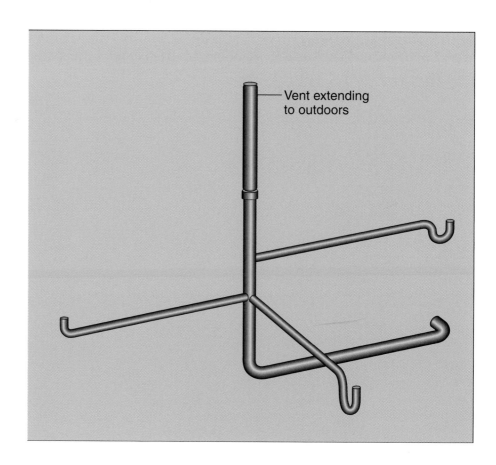

Vent extending to outdoors

P3102.1 Required Vent Extension. The vent system serving each building drain shall have at least one vent pipe that extends to the outdoors.

P3102.2 Installation. The required vent shall be a dry vent that connects to the building drain or an extension of a drain that connects to the building drain. Such vent shall not be an island fixture vent as allowed by Section P3112.

P3102.3 Size. The required vent shall be sized in accordance with Section P3113.1 based on the required size of the building drain.

CHANGE SIGNIFICANCE. The 2003 IRC Section P3102.1 contains several areas that needed improvement. For example, this section states that every building shall have either a stack vent or a vent stack, even though it is possible in cases such as slab on grade to have no waste stack (only connections to the building drain). There are also several terms used that are unclear and could cause incorrect application of the code. Thus, the term "undiminished" is sometimes misapplied to mean undiminished size of the building drain and not the vent, and "directly as possible" does not provide much guidance for offset of a vent through the roof extending out the back to avoid visible penetrations at the front portion of a roof. Additionally, the 2003 IRC Section P3102.2 addresses vent stack installations but does not provide information about when a vent stack is required. Because the IRC is limited in scope to certain residential buildings three stories or less in height, a vent stack is never required in an IRC structure; plumbing code typically require vent stacks every five or more branch intervals. To address such ambiguities and redundancies, the 2006 code now has introduced new language covering the needed information in a concise manner and has eliminated all the other unnecessary information.

P3103.1
Roof Extension

CHANGE TYPE.   Modification

CHANGE SUMMARY.   Vent pipes through the roof are now required to be extended a minimum of 6 inches above the roof. In geographic areas subject to heavy snow, vents must extend at least 6 inches above the anticipated snow accumulation.

2006 CODE: P3103.1 Roof Extension.   Open vent pipes that extend through a roof shall be terminated at least <u>6</u> [number] inches <u>(152 mm)</u> above the roof or [number] <u>6</u> inches <u>(152 mm)</u> above the anticipated snow accumulation, <u>whichever is greater,</u> except that where a roof is to be used for any purpose other than weather protection, the vent extension shall be run at least 7 feet (2134 mm) above the roof.

CHANGE SIGNIFICANCE.   The 2003 code leaves the decision regarding the length of extension for vent pipes above the roof in the hands of each jurisdiction. This is because there might be special circumstances such as deep snow on roofs that could not be addressed nationally and must be decided in each jurisdiction. The main intent, of course, is that the vent pipes extend high enough to allow for roof flashing and not be obstructed. This approach has caused nonuniformity across the country with regard to this subject and has forced each and every jurisdiction adopting the IRC to go through an independent process of deciding an extension length. In light of the fact that uniformity is one of the main objectives of model codes, the 2006 code has been modified to require a specific measure, 6 inches, of vent pipe extension above the roof. If snow accumulation is expected, the vent pipe must extend to 6 inches above the anticipated snow accumulation level.

P3105.2
Fixture Drains

CHANGE TYPE. Deletion

CHANGE SUMMARY. The methodology allowing a certain type of "S" trap, designated in the code as "vertical leg for waste fixture drains" and found in the 2003 code Section P3105.3, has been deleted. The two references to this methodology from Sections P3105.2 and P3201.5 have also been deleted.

2006 CODE: P3105.2 Fixture Drains. The total fall in a fixture drain due to pipe slope shall not exceed one pipe diameter, nor shall the vent pipe connection to a fixture drain, except for water closets, be below the weir of the trap.~~, except as provided in Section P3105.3.~~

~~**P3105.3 Vertical Leg for Waste Fixture Drains.** A vertical leg (see Figure P3105.3) is permitted within a fixture drain of a waste fixture in accordance with the following criteria:~~

~~1. Minimum trap diameter shall be in accordance with Table P3201.7.~~
~~2. The diameter of Section A shall be equal to the diameter of the trap.~~
~~3. The length of Section A shall not be less than 8 inches (203 mm) and in accordance with Table P3105.1.~~
~~4. The diameter of Section B shall be one pipe size larger than the diameter of Section A.~~
~~5. The length of Section B shall not be more than 36 inches (914 mm).~~
~~6. The diameter of Section C shall be one pipe size larger than the diameter of Section B.~~

P3105.2 continues

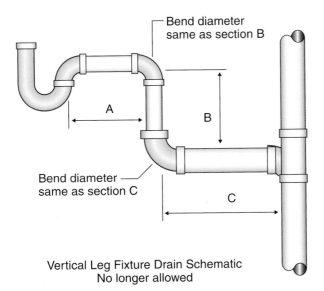

Vertical Leg Fixture Drain Schematic
No longer allowed

P3105.2 continued

7. ~~The total length of Section A and Section C shall not exceed the distance allowed in Table P3105.1.~~

8. ~~Bends shall be the diameter of the largest connected section.~~

~~**FIGURE P3105.3**   **Vertical Leg Fixture Drain Schematic**~~

P3201.5 Prohibited Trap Designs.   The following types of traps are prohibited:

1. Bell traps.
2. Separate fixture traps with interior partitions, except those lavatory traps made of plastic, stainless steel, or other corrosion resistant material.
3. "S" traps ~~(except as permitted under Section P3105.3)~~.
4. Drum traps.
5. Trap designs with moving parts.

CHANGE SIGNIFICANCE.   The configuration described and shown in the related figure of 2003 code Section P3105.3 creates a form of an S-trap, and it does not appear to have any justification for being allowed in the code. A combination drain and vent system is permitted under Section 3111 and has proved to be a useful venting method for this type of configuration. It is believed that the "vertical leg" is susceptible to far greater risk of trap seal loss than other venting methods and therefore should be avoided. These are the reasons why the IPC does not allow this method of venting. Furthermore, this exception to the long-standing prohibition of "S" trap configurations, found in Section P3201.5, is unnecessary when an air-admittance-valve venting system can provide yet another option.

CHANGE TYPE. Clarification

CHANGE SUMMARY. The figures referenced by this section have been deleted from the body of the code and relocated to a new Appendix N, venting methods, to eliminate conflicts and confusion between the code text and the figures. Additionally, new language has been introduced clarifying that only horizontally connected fixtures are allowed to connect to a horizontal wet vent.

2006 CODE: P3108.1 <u>Horizontal</u> Wet Vent Permitted. Any combination of fixtures within two bathroom groups located on the same floor level are permitted to be vented by a <u>horizontal</u> wet vent. The wet vent shall be considered the vent for the fixtures and shall extend from the connection of the dry vent along the direction of the flow in the drain pipe to the most downstream fixture drain connection. <u>Each fixture drain shall connect horizontally to the horizontal branch being wet vented or shall have a dry vent.</u> Only the fixtures within the bathroom groups shall connect to the wet vented horizontal branch drain. Any additional fixtures shall discharge downstream of the <u>horizontal</u> wet vent. [~~See Figures P3108.1(1), P3108.1(2) and~~ ~~P3108.1(3) for typical wet vent configurations.~~]

~~**FIGURE P3108.1(1)**~~   ~~**Typical Single Bath Wet Vent Arrangements**~~

~~**FIGURE P3108.1(2)**~~   ~~**Typical Double Bath Wet Vent Arrangements**~~

~~**FIGURE P3108.1(3)**~~   ~~**Typical Horizontal Wet Venting**~~

P3108.1 continues

P3108.1
Wet Vent Permitted

Example of Horizontal Wet Venting

P3108.1 continued

APPENDIX N: VENTING METHODS

(This appendix is informative and is not part of the code. This appendix provides examples of various illustrations of venting methods.)

FIGURE N1   **Typical Single-Bath Wet-Vent Arrangements**

FIGURE N2   **Typical Double-Bath Wet-Vent Arrangements**

FIGURE N3   **Typical Horizontal Wet Venting**

(Four additional figures are provided in this appendix.)

CHANGE SIGNIFICANCE. The illustrations provided in the 2003 code Section P3108.1 are intended as a guide. In many cases, however, these figures have helped create installation and enforcement problems, as some believe the methods shown are the only methods permitted to be used for these installations, including exact fitting location and order shown. As these figures are intended to be guideline illustrations, and the code body should not contain guidelines, the 2006 code has deleted these figures from Chapter 31 and relocated them to a new Appendix N, where they will still serve their purpose as a guide. The same change has occurred in other sections of the code such as Sections P3108.4, P3109.2, and P3110.4, where figures in the code have created similar problems. The figures in these sections have also been deleted from Chapter 31 and relocated to the new Appendix N.

Another change in Section P3108.1 is the insertion of new language to clarify the original intent of this section, that only horizontally connected fixtures are allowed to connect to a horizontal wet vent. Some jurisdictions allow fixtures to rise vertically from the horizontal wet vent and then travel horizontally to the fixture. This creates an S-trap because the horizontal fixture drain is now above the vent. Without the new language in the 2006 code, a configuration that was never intended by the code may be installed.

P3108.2 and P3108.3

Vent Connections (P3108.2) and Size (P3108.3)

CHANGE TYPE.   Clarification

CHANGE SUMMARY.   New language has been added to the text to clarify the allowed arrangement of fixtures in wet venting systems and how to size the dry vent serving the wet vent.

2006 CODE: P3108.2 Vent Connections.   The dry vent connection to the wet vent shall be an individual vent or common vent to the lavatory, bidet, shower, or bathtub. In vertical wet vent systems the most upstream fixture drain connection shall be a dry-vented fixture drain connection. In horizontal wet vent systems, not more than one wet vented fixture drain shall discharge upstream of the dry-vented fixture drain connection.

P3108.3 Size.   Horizontal and vertical wet vents shall be of a minimum size as specified in Table P3108.3, based on the fixture unit discharge to the wet vent. The dry vent serving the wet vent shall be sized based on the largest required diameter of pipe within the wet vent system served by the dry vent.

CHANGE SIGNIFICANCE.   The new language in Section P3108.2 clarifies and simply acknowledges the fact that in vertical wet-vent systems the point where the dry vent connects to the wet vent is always located at the most upstream fixture drain connection to the vertical stack. It also makes it clear that in horizontal wet-vent systems, multiple wet-vented fixtures within the bathroom groups are not permitted to discharge upstream of the fixture providing the vent for the

P3108.2 and P3108.3 continues

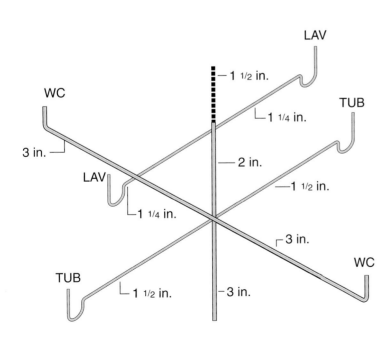

LAV

1 1/2 in.

WC

TUB

1 1/4 in.

3 in.

2 in.

LAV

1 1/2 in.

1 1/4 in.

3 in.

WC

TUB

1 1/2 in.

3 in.

Example of Vertical Wet Venting

P3108.2 and P3108.3 continued

system. When horizontal wet vents are installed in this manner, fixture traps upstream of the dry-vented fixture drain connection can be subjected to negative pressures. Also, the 2003 code does not address the sizing criteria for the dry vent serving the wet vent. New text in 2006 code Section P3108.3 has been added to require that such dry vents be sized on the basis of the largest required diameter of pipe within the wet-vent system.

CHANGE TYPE. Modification

P3201.6

Number of Fixtures per Trap

CHANGE SUMMARY. The maximum vertical distance from the fixture outlet to the trap weir is required in the 2003 code to be 24 inches, but a horizontal distance is not specified, other than that the trap must be located as close as possible to the fixture outlet. The 2006 code includes the 24-inch maximum vertical distance and adds a provision for the maximum horizontal distance to be 30 inches.

2006 CODE: P3201.6 Number of Fixtures Per Trap.   Each plumbing fixture shall be separately trapped by a water seal trap ~~placed as close as possible to the fixture outlet.~~ The vertical distance from the fixture outlet to the trap weir shall not exceed 24 inches (610 mm) <u>and the horizontal distance shall not exceed 30 inches (762 mm) measured from the center line of the fixture outlet to the centerline of the inlet of the trap.</u> The ~~distance~~ <u>height</u> of a clothes washer standpipe above a trap shall conform to Section P2706.2. Fixtures shall not be double trapped.

Exceptions:
1. Fixtures that have integral traps.
2. A single trap shall be permitted to serve two or three like fixtures limited to kitchen sinks, laundry tubs, and lavatories. Such fixtures shall be adjacent to each other and located in the same room with a continuous waste arrangement. The trap shall be installed at the center fixture where three such fixtures are installed. Common trapped fixture outlets shall not be more than 30 inches (762 mm) apart.
3. Connection of a laundry tray waste line into a standpipe for the automatic clothes-washer drain is permitted in accordance with Section P2706.2.1.

CHANGE SIGNIFICANCE. The vertical and horizontal dimensions from the fixture outlet to the trap seal are a function of the energy needed to keep the water seal intact while allowing waste to move through the trap. The 2003 code language "placed as close as possible

P3201.6 continues

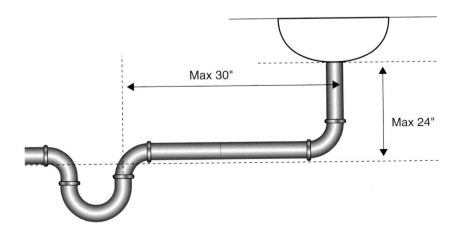

P3201.6 continued   to the fixture outlet," while subjective, allows for what many believe to be flexible criteria based on good judgment and for which there are no documented problems. The 2006 code has now deleted this subjective language and inserted a maximum 30-inch horizontal distance from the fixture outlet to the trap inlet centerline, similar to what is found under exception 2 for three compartment sinks. This dimension is believed to be the maximum length through which the flow energy is able to move waste. Although this change will appear in the 2006 code, there is not a consensus on this subject among the professionals and the industry. Many believe the 30-inch horizontal distance creates a running trap for which there is not enough energy to move waste through and that a system designed with such horizontal length will be subject to stoppage. Other major plumbing codes do not contain such an allowance or maximum length for horizontal distance.

CHANGE TYPE.   Modification

CHANGE SUMMARY.   Shower trap and trap arm sizes have been revised to account for various flow rates in one shower.

2006 CODE: Table P3201.7 Size of Traps and Trap Arms for Plumbing Fixtures (see page 266).

CHANGE SIGNIFICANCE.   There are many cases where multiple shower heads and body sprays are used in showers. The 2003 code, in its table for shower trap and trap arm sizes, does not account for these situations. There have been documented cases in which 17 showerheads and body sprays have been installed in a single shower. As it has been necessary to account for such multiple shower heads and body sprays and situations which produce higher flow rates, the 2006 code has attempted to address this issue by inserting new information into Table P3201.7. The proponent of the proposed code change number P91-04/05 performed extensive calculations for numerous conditions with various numbers of shower heads and various friction factors for common pipes and arrived at the minimum trap sizes shown as new text in the 2006 code.

Table P3201.7 continues

Table P3201.7

Size of Traps and Trap Arms for Plumbing Fixtures

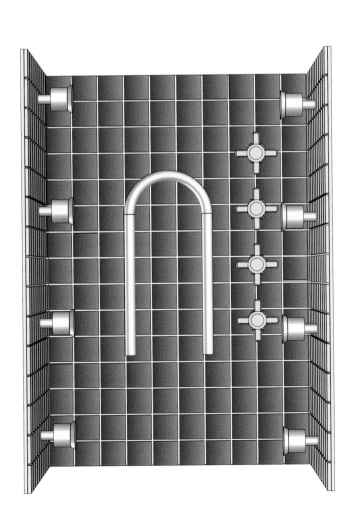

Table P3201.7 continued

TABLE P3201.7   Size of Traps and Trap Arms for Plumbing Fixtures

Plumbing Fixture	Trap Size Minimum (inches)
Bathtub (with or without shower head and/or whirlpool attachments)	1½
Bidet	1¼
Clothes washer standpipe	2
Dishwasher (on separate trap)	1½
Floor drain	2
Kitchen sink (one or two traps, with or without dishwasher and garbage grinder)	1½
Laundry tub (one or more compartments)	1½
Lavatory	1¼
~~Shower~~	~~2~~
Shower (based on the total flow rate through showerheads and bodysprays) Flow rate:	
5.7 g.p.m. or less	1-½"
More than 5.7 g.p.m. to 12.3 g.p.m.	2"
More than 12.3 g.p.m. to 25.8 g.p.m.	3"
More than 25.8 g.p.m. to 55.6 g.p.m.	4"
Water closet	Note a

For SI: 1 inch = 25.4 mm.

a. Consult fixture standards for trap dimensions of specific bowls.

CHANGE TYPE. Modification

CHANGE SUMMARY. The circuit identification requirements for panelboards have been expanded to provide for a more detailed representation of the rooms, areas, or equipment being served by each circuit.

2006 CODE: E3606.2 Panelboard Circuit Identification. All circuits and circuit modifications shall be legibly identified as to <u>their clear, evident, and specific</u> purpose or use. <u>The identification shall include sufficient detail to allow each circuit to be distinguished from all others. The identification shall be included in</u> ~~on~~ a circuit directory located on the face <u>of the panelboard enclosure</u> or inside the <u>panel</u> door ~~of the enclosure~~.

CHANGE SIGNIFICANCE. The additional language addressing panelboard circuit identification emphasizes the importance of detailed information. Although the labeling of panelboards has long been a code requirement, no direction was previously given as to the amount of information that needed to be provided. There will continue to be a need for the appropriate interpretation of "clear, evident, and specific" labels, but it is obvious that generic terminology is no longer appropriate.

E3606.2
Panelboard Circuit Identification

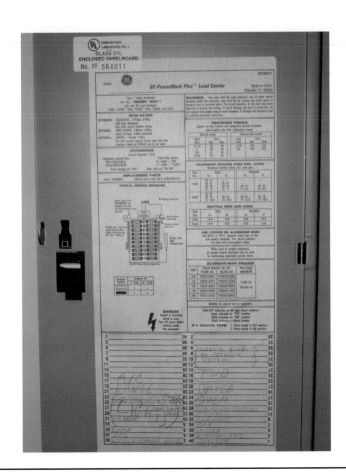

E3702.2.2, Table E3702.1

Protection of Cables Parallel to Furring Strips

CHANGE TYPE.   Addition

CHANGE SUMMARY.   Concern about damage to cables installed parallel to furring strips is now addressed by requiring such cables to be protected in the same manner as cables installed adjacent to framing members.

2006 CODE: E3702.2.2 Cable Installed Parallel to Framing Members.   Where cables are installed <u>through or</u> parallel to the sides of rafters, studs or floor joists, guard strips and running boards shall not be required, and the installation shall comply with Table E3702.1.

TABLE E3702.1   **General Installation and Support Requirements for Wiring Methods**[a,b,c,d,e,f,g,h,i,j,k]

Installation Requirements (Requirement applicable only to wiring methods marked "A")	AC MC	EMT IMC RMC	ENT	FMC LFC	NM UF	RNC	SE	SR[a]	USE
Where run parallel with the framing member <u>or furring strip,</u> the wiring shall be <u>not less than</u> ~~1.25~~ 1¼ inches from the edge of <u>a furring strip or</u> a framing member such as a joist, rafter, or stud or shall be physically protected.	A	—	A	A	A	—	A	—	—

(no significant changes to remainder of Table)

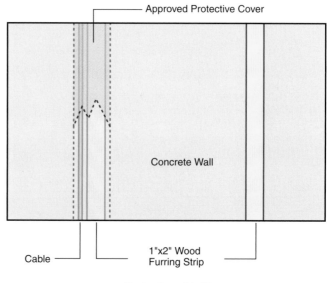

Protection of Cables

Reproduced from material prepared by the NFPA based on the 2005 National Electrical Code, as published in the 2006 International Residential Code, Copyright © 2005, National Fire Protection Association.

CHANGE SIGNIFICANCE. The hazards and concerns related to cables installed parallel to framing members such as floor and ceiling joists, rafters, and studs are also pertinent when furring strips are utilized for fastening purposes. For this reason, furring strips are now regulated in the same manner as such framing members. Because of the potential for damage to cables adjacent to framing members and furring strips, Table E3702.1 requires a minimum $1\frac{1}{4}$-inch separation between the cable installation and the edge of the wood member. Where the $1\frac{1}{4}$-inch clearance cannot be met, as is common with shallow furring strips, an approved method of physical protection is acceptable.

Although framing members are often provided for structural support, they are also typically utilized for fastening purposes. When used in this manner, they are no different than furring strips also utilized as the support for plywood sheathing or gypsum wallboard. Because there is a significant chance that a fastener that is intended for penetration into a furring strip will instead penetrate a cable, an adequate clearance or appropriate physical protection must be provided.

E3702.3.2

Protecting Type NM Cable from Physical Damage

CHANGE TYPE.   Modification

CHANGE SUMMARY.   The use of pipe or guard strips to protect Type NM cable from physical damage is no longer permitted. In addition, the type of conduit permitted as a protective sleeve is specifically identified as rigid metal or intermediate metal.

2006 CODE: E3702.3.2 Protection from Physical Damage. Where subject to physical damage, cables shall be protected by <u>rigid metal</u> conduit, <u>intermediate metal conduit,</u> electrical metallic tubing, Schedule 80 PVC rigid nonmetallic conduit~~, pipe, guard strips~~ or other approved means. Where passing through a floor, the cable shall be enclosed in rigid metal conduit, intermediate metal conduit, electrical metallic tubing, Schedule 80 PVC rigid nonmetallic conduit, or other ~~metal pipe~~ <u>approved means,</u> extending not less than 6 inches (152 mm) above the floor.

CHANGE SIGNIFICANCE.   Type NM cable must be protected when it is installed in locations subject to physical damage. It has been deemed that pipe and guard strips do not reach the level of necessary protection as that provided by rigid metal conduit, intermediate metal conduit, electrical metallic tubing, and Schedule 80 PVC rigid nonmetallic conduit. Therefore, the use of such materials for protection purposes is now prohibited.

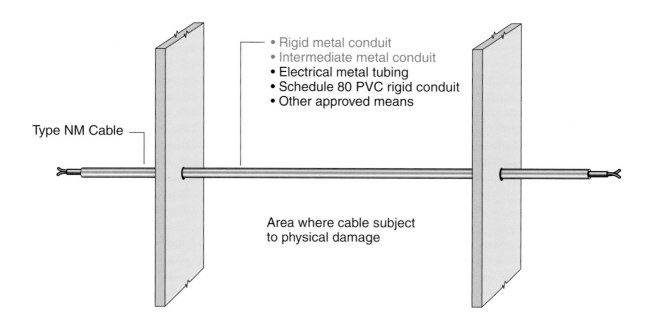

Protecting Type NM Cable from Physical Damage

CHANGE TYPE.   Addition

CHANGE SUMMARY.   Covering with a listed "sunlight-resistant" insulating material is now an acceptable method of protecting conductors directly exposed to sunlight.

2006 CODE: E3702.3.3 Locations Exposed to Direct Sunlight. Insulated conductors and cables used where exposed to direct rays of the sun shall be of a type listed for sunlight resistance, or of a type listed and marked "sunlight resistant," or shall be covered with insulating material, such as tape or sleeving, that is listed or listed and marked as being "sunlight resistant."

CHANGE SIGNIFICANCE.   Because of the damaging effects of direct sunlight on insulated conductors and cables, they must be recognized as a type listed for sunlight resistance. As an alternative, the code now specifically permits the use of a listed tape or sleeving material as a means of protecting the conductors and cables from the sun's harmful rays.

E3702.3.3

Conductors and Cables Exposed to Direct Sunlight

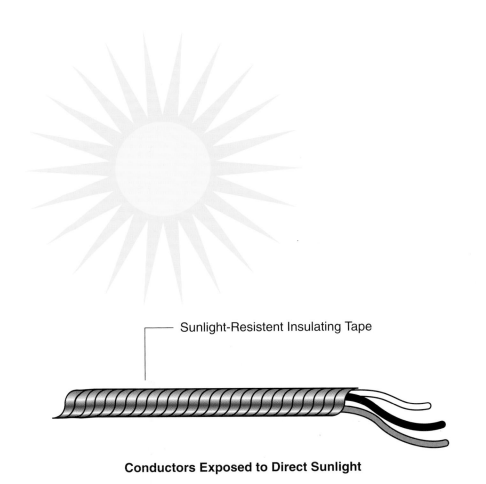

Sunlight-Resistent Insulating Tape

Conductors Exposed to Direct Sunlight

E3801.3

Small-Appliance Circuit Receptacle Outlets

CHANGE TYPE.   Clarification

CHANGE SUMMARY.   Where a minimum of two appliance branch circuits are required to serve receptacle outlets in the kitchen, dining room, and similar areas, all receptacle outlets, including those installed in the floors of such rooms and areas, are regulated.

2006 CODE: E3801.3 Small Appliance Receptacles.   In the kitchen, pantry, breakfast room, dining room, or similar area of a dwelling unit, the two or more 20-ampere small-appliance branch circuits required by Section E3603.2, shall serve all <u>wall and floor</u> receptacle outlets covered by Sections E3801.2 and E3801.4 and those receptacle outlets provided for refrigeration appliances.
(no change to exceptions)

CHANGE SIGNIFICANCE.   In a dwelling unit, the kitchen, dining room, and other similar areas must be served by at least two small-

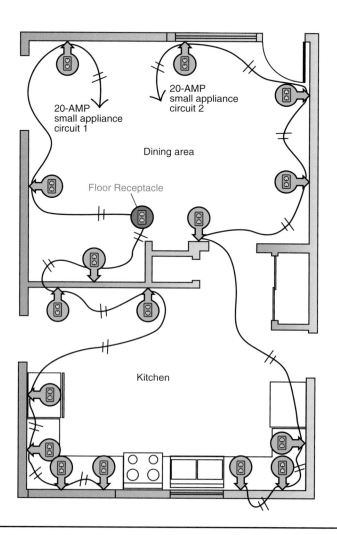

appliance branch circuits. By specifically addressing both wall outlets and floor outlets, the code change has clarified the application of the provisions. It is not uncommon for dining areas to contain floor outlets, and the code now specifies that such outlets shall be served by one of the required small-appliance circuits.

E3801.4.1, E3801.4.2

Receptacle Outlets at Wall and Island Counter Spaces

CHANGE TYPE.   Modification

CHANGE SUMMARY.   The method for determining required locations for receptacle outlets at counter spaces has been clearly defined for those wall areas adjacent to a sink or range. In addition, the required outlet locations at island counter spaces have been clarified where a rangetop or sink is installed.

2006 CODE: E3801.4.1 Wall Counter Space.   A receptacle outlet shall be installed at each wall counter space 12 inches (305 mm) or wider. Receptacle outlets shall be installed so that no point along the wall line is more than 24 inches (610 mm), measured horizontally from a receptacle outlet in that space.

> **Exception:** Receptacle outlets shall not be required on a wall directly behind a range or sink in the installation described in Figure E3801.4.1.

E3801.4.2 Island Counter Spaces.   At least one receptacle outlet shall be installed at each island counter space with a long dimension of 24 inches (610 mm) or greater and a short dimension of 12 inches (305 mm) or greater. Where a rangetop or sink is installed in an island counter and the width of the counter behind the rangetop or sink is less than 12 inches (300 mm), the rangetop or sink has divided the island into two separate countertop spaces as defined in Section E3801.4.4.

CHANGE SIGNIFICANCE.   Figure E3801.4.1 has been added to the code to specifically illustrate the exempt locations for wall receptacle outlets at countertops. Previously, varied interpretations of the code language resulted in the inconsistent application of the provisions. Two different installation locations are addressed in the figure, where the sink or range extends from the face of the counter, and where the

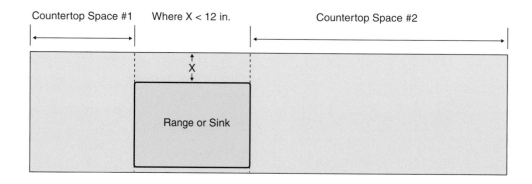

Receptacles at Island Counter Spaces

sink or range is mounted in a corner. It is important to note that both conditions are based on a limited countertop depth behind the sink or range.

The previous language regulating receptacle outlets in island counter spaces has also been expanded to specifically identify when a rangetop or sink divides the island into two separate countertop spaces. Such division occurs only where the depth of the counter behind the range top or sink is less than 12 inches. In those situations where the depth equals or exceeds 12 inches, the island countertop shall be considered as a single counter space. This distinction could be critical when determining the required number and location of receptacle outlets.

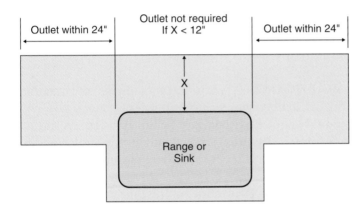

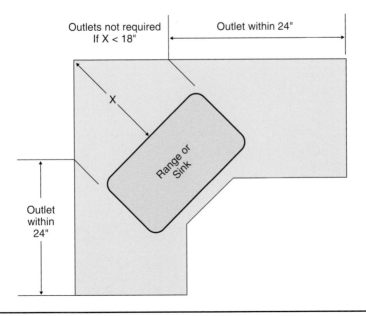

E3801.6

Bathroom Receptacles

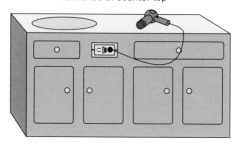

Outlet with 3 feet of
basin edge and within
12 inches of counter top

CHANGE TYPE. Modification

CHANGE SUMMARY. The required bathroom receptacle outlet can now be located on the side or face of the lavatory basin cabinet, provided it is within 12 inches of the countertop surface, measured vertically.

2006 CODE: E3801.6 Bathroom. At least one wall receptacle outlet shall be installed in bathrooms and such outlet shall be located within 36 inches (914 mm) of the outside edge of each lavatory basin. The receptacle outlet shall be located on a wall that is adjacent to the lavatory basin location.

Receptacle outlets shall not be installed in a face-up position in the work surfaces or countertops in a bathroom basin location.

> **Exception:** <u>The receptacle shall not be required to be mounted on the wall or partition where it is installed on the side or face of the basin cabinet not more than 12 inches (300 mm) below the countertop.</u>

CHANGE SIGNIFICANCE. A minimum of at least one receptacle outlet is required in all bathrooms. Such outlet must be located within 3 feet of the outside edge of each lavatory basin. Previously, the location for the receptacle outlet was limited to a wall surface adjacent to the basin. The new exception permits the required outlet to be installed on the side or face of the basin cabinet rather than on a wall or partition surface, but only where it is located in close proximity to the top of the basin. Where the outlet is mounted on the face or side of the lavatory basin, its location must comply with both the 36-inch limitation on the distance to the lavatory edge and the 12-inch limitation on the distance below the countertop.

CHANGE TYPE.   Modification

CHANGE SUMMARY.   A receptacle outlet is no longer required to be installed adjacent to an evaporative cooler for the purpose of servicing the unit.

2006 CODE: E3801.11 HVAC Outlet.   A 125-volt, single-phase, 15 or 20 ampere-rated convenience receptacle outlet shall be installed for the servicing of heating, air-conditioning, and refrigeration equipment located in attics and crawl spaces. The receptacle shall be accessible and shall be located on the same level and within 25 feet (7620 mm) of the heating, air-conditioning, and refrigeration equipment. The receptacle outlet shall not be connected to the load side of the HVAC equipment disconnecting means. ~~and shall be protected in accordance with Section E3802.4.~~

> **Exception:** <u>A receptacle outlet shall not be required for the servicing of evaporative coolers.</u>

E3801.11 continues

E3801.11

Heating, Air-Conditioning, and Refrigeration Equipment Outlet

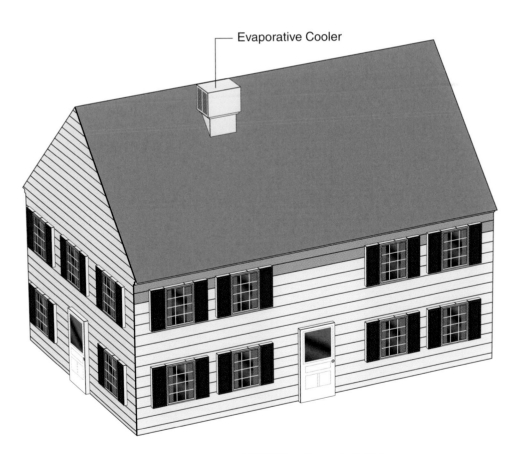

HVAC Equipment Outlets

E3801.11 continued

CHANGE SIGNIFICANCE.   Typically, a receptacle outlet must be provided near any heating, air-conditioning, or refrigeration equipment. The outlet assists in the efficient and safe servicing of the equipment by providing an electrical connection for tools and other equipment necessary in the maintenance process. A new exception exempts evaporative coolers from the outlet requirement. It was determined that the needs for the servicing, maintenance, and replacement of evaporative coolers are not as extensive as those for other types of air-conditioning equipment.

E3802.7

Ground-Fault Circuit-Interrupter Protection at Laundry and Utility Sinks

CHANGE TYPE. Addition

CHANGE SUMMARY. Receptacles located adjacent to a wet bar sink must now also be provided with ground-fault circuit-interrupter (GFCI) protection where the receptacles do not serve a countertop surface. In addition, laundry sinks and utility sinks are now regulated in the same manner as wet bar sinks.

2006 CODE: E3802.7 <u>Laundry, Utility, and</u> B<u>Bar Sink Recep-</u>tacles. All 125-volt, single-phase, 15- and 20-ampere receptacles that ~~serve a countertop surface, and~~ are located within 6 feet (1829 mm) of the outside edge of a <u>laundry, utility, or</u> wet bar sink shall have ground-fault circuit-interrupter protection for personnel. Receptacle outlets shall not be installed in a face-up position in the work surfaces or countertops.

CHANGE SIGNIFICANCE. Previously, only those receptacles that served a countertop surface in close proximity to a wet bar sink were required to be GFCI-protected. The code change eliminated the refer-

E3802.7 continues

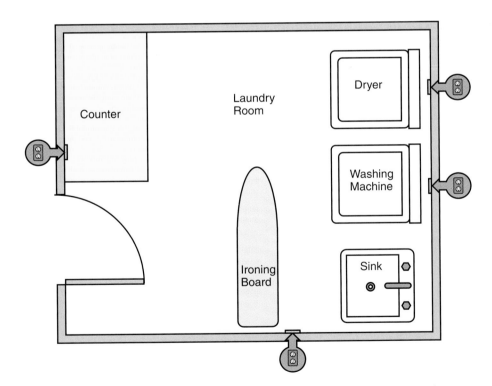

All receptacles within 6 feet of outside edge of sink to be GFCI-protected

GFCI Protection at Laundry Sinks

E3802.7 continued   ence to countertop surfaces, expanding the application of the provision to those locations without countertops. The 6-foot measurement continues to be taken from the outside edge of the sink to the receptacle location. The scope of the provision was also expanded by regulating receptacles within 6 feet of laundry sinks and utility sinks.

CHANGE TYPE.   Modification

CHANGE SUMMARY.   With limitations, arc-fault circuit interrupters (AFCIs) may now be in locations other than the origination of the branch circuit. Whereas branch/feeder-type AFCI protection continues to be permitted for branch circuits serving bedrooms, only combination-type devices will be acceptable beginning January 1, 2008.

2006 CODE: E3802.~~11~~12 Arc-Fault Protection of Bedroom Outlets.   All branch circuits that supply 125-volt, single-phase 15- and 20-ampere outlets installed in dwelling unit bedrooms shall be protected by ~~an~~ a combination type or branch/feeder type arc-fault circuit interrupter ~~listed~~ installed to provide protection of the entire branch circuit. Effective January 1, 2008, such arc-fault circuit interrupter devices shall be combination type.

E3802.12 continues

E3802.12
Arc-Fault Circuit-Interrupter Protection in Bedrooms

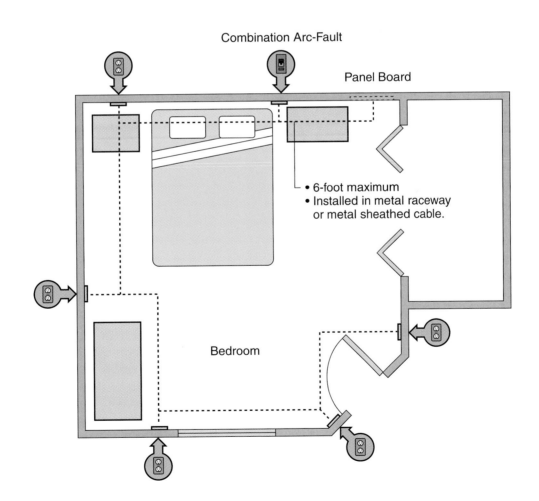

AFCI-Protection in Bedrooms

E3802.12 continued

Exception: <u>The location of the arc-fault circuit interrupter shall be permitted to be at other than the origination of the branch circuit, provided that:</u>

1. <u>The arc-fault circuit interrupter is installed within 6 feet (1.8 m) of the branch circuit overcurrent device as measured along the branch circuit conductors, and</u>
2. <u>The circuit conductors between the branch circuit overcurrent device and the arc-fault circuit interrupter are installed in a metal raceway or a cable with a metallic sheath.</u>

CHANGE SIGNIFICANCE.   The preferred type of AFCI is the combination-type device, but such AFCI devices are not mandated until January 1, 2008. Until that time, the use of branch/feeder-type devices is also acceptable. Prior to 2008, either type of AFCI device may be utilized for the protection of branch circuits supplying bedroom outlets.

Branch/feeder-type devices are intended to be installed at the point of origin of a branch circuit or feeder, whereas the requirements for both branch/feeder and outlet circuit AFCIs are met with combination-type devices. The new exception permits the AFCI to be located at a point other than that of the branch circuit origin, but only where specified limitations are met.

CHANGE TYPE. Addition

CHANGE SUMMARY. Screws are now specifically recognized as an acceptable means of supporting boxes; however, any screw threads exposed within a box must be protected to eliminate damage due to abrasion.

2006 CODE: E3806.8.2.1 Nails and Screws. Nails and screws, where used as a fastening means, shall be attached by using brackets on the outside of the enclosure, or they shall pass through the interior within 0.25 ¼ inch (6.4 mm) of the back or ends of the enclosure. Screws shall not be permitted to pass through the box except where exposed threads in the box are protected by an approved means to avoid abrasion of conductor insulation.

CHANGE SIGNIFICANCE. Nails have traditionally been recognized in the code as the sole acceptable attachment for the support of boxes and enclosures. Screws are now also considered as an appropriate fastening means. A concern with the use of screws is the potential for damage to conductor insulation caused by abrasive action. Therefore, it is necessary that the exposed threads within the box or enclosure be adequately protected.

E3806.8.2.1
Enclosure Support Using Nails and Screws

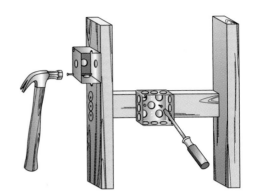

E3807.2

Enclosures in Wet Locations

CHANGE TYPE.   Addition

CHANGE SUMMARY.   Where cabinets and panelboards are installed in damp or wet locations, fittings for raceways and cables entering above the level of uninsulated live parts must be listed for use in wet locations.

2006 CODE: E3807.2 Damp or Wet Locations.   In damp or wet locations, cabinets and panelboards of the surface type shall be placed or equipped so as to prevent moisture or water from entering and accumulating within the cabinet, and shall be mounted to provide an airspace not less than ~~0.25~~ ¼ inch (6.4 mm) between the enclosure and the wall or other supporting surface. Cabinets installed in wet locations shall be weatherproof. For enclosures in wet locations, raceways and cables entering above the level of uninsulated live parts shall be installed with fittings listed for wet locations.

CHANGE SIGNIFICANCE.   When installed in wet locations, surface-type cabinets and panelboards must be weatherproof and placed in a manner to prevent moisture from entering the cabinet. A new provision requires that any raceways or cables that enter above the level of uninsulated live parts be provided with fittings that are listed for use in wet locations.

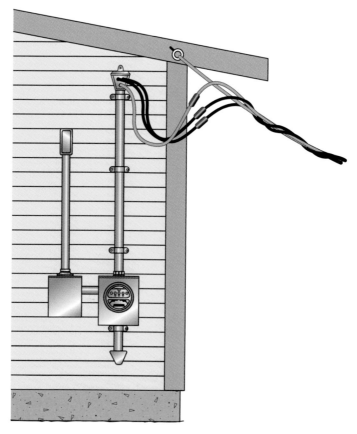

Reproduced from material prepared by the NFPA based on the 2005 National Electrical Code, as published in the 2006 International Residential Code, Copyright © 2005, National Fire Protection Association.

CHANGE TYPE. Addition

CHANGE SUMMARY. A provision mandating the repair of damaged or incomplete drywall and plaster surfaces adjacent to panelboards has been added, the requirements of which are consistent with those of Section E3806.6 for boxes.

2006 CODE: <u>**E3807.4 Repairing Plaster, Drywall and Plaster-board.**</u> <u>Plaster, drywall, and plasterboard surfaces that are broken or incomplete shall be repaired so that there will not be gaps or open spaces greater than ⅛ inch (3.2 mm) at the edge of the cabinet or cutout box employing a flush-type cover.</u>

CHANGE SIGNIFICANCE. Section E3806.6 requires that openings in plaster, drywall, or plasterboard surfaces intended to accommodate

E3807.4 continues

Reproduced from material prepared by the NFPA based on the 2005 National Electrical Code, as published in the 2006 International Residential Code, Copyright © 2005, National Fire Protection Association.

E3807.4
Repairing Drywall at Panelboards

E3807.4 continued boxes be made such that there are no gaps or open spaces greater than $\frac{1}{8}$ inch around the edge of the box. Any panelboard with a flush-type cover is now regulated in a similar manner. Quite often, slightly oversized gaps and damaged edges can be repaired with drywall joint compound.

CHANGE TYPE.   Addition

CHANGE SUMMARY.   Receptacles are now prohibited directly over a bathtub or shower stall, as well as within the tub or shower area. In addition, weatherproof faceplate assemblies are mandated for flush-mounted outlet boxes in all finished surfaces, not just walls, where located in damp or wet locations.

2006 CODE: **E3902.~~10~~11 Bathtub and Shower Space.**   A receptacle shall not be installed within <u>or directly over</u> a bathtub or shower ~~space~~ <u>stall</u>.

E3902.~~11~~12 Flush Mounting With Faceplate.   In damp or wet locations, the enclosure for a receptacle installed in an outlet box flush-mounted ~~on~~ <u>in</u> a ~~wall~~ <u>finished</u> surface shall be made weatherproof by means of a weatherproof faceplate assembly that provides a water-tight connection between the plate and the ~~wall~~ <u>finished</u> surface.

CHANGE SIGNIFICANCE.   The prohibition of receptacles within bathtub or shower areas was often applied only to walls surrounding the tub or shower area. The new language specifically identifies the area directly above a bathtub or shower stall as an additional location where receptacles are prohibited. In a code change of similar application, flush-mounted outlet boxes in all finished surfaces located in damp or wet locations must be provided with weatherproof faceplate assemblies. Previously, the requirement for weatherproofing was limited to outlet boxes mounted in walls.

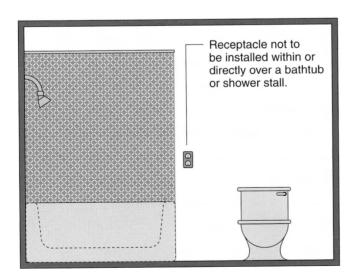

Receptacle not to be installed within or directly over a bathtub or shower stall.

Receptacles in Wet Locations

E3903.10

Luminaires in Bathtub and Shower Areas

CHANGE TYPE.   Modification

CHANGE SUMMARY.   Specific types of hanging luminaires are now regulated as to their location when in close proximity to bathtubs and shower stalls. Luminaires located in these tub and shower areas must be listed for damp locations, with a wet-location listing required where subject to shower spray.

2006 CODE: E3903.10 Bathtub and Shower Areas. Cord-connected luminaires, ~~hanging~~ chain-, cable-, or cord-suspended luminaires, lighting track, pendants, and ceiling-suspended (paddle) fans shall not have any parts located within a zone measured 3 feet (914 mm) horizontally and 8 feet (2438 mm) vertically from the top of a bathtub rim or shower stall threshold. The zone is all-encompassing and includes the zone directly over the tub or shower. Luminaires located in this zone shall be listed for damp locations, and where subject to shower spray, shall be listed for wet locations.

CHANGE SIGNIFICANCE.   The code identifies a zone directly above and adjacent to bathtubs and shower stalls where specified luminaires, lighting track, pendants, and ceiling fans are prohibited. Regarding the prohibition on specific luminaires, the generic term "hanging" has been replaced with "chain-, cable-, or cord-suspended"

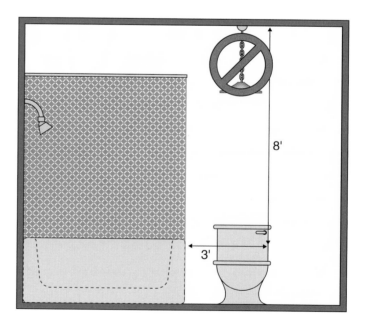

Luminaires in Bathtub and Shower

in order to clarify that the provisions are not applicable to approved luminaires mounted on a wall.

Although specific luminaires are prohibited in the tub and shower zone, other types, such as those that are surface-mounted, are acceptable when listed for damp locations. A listing recognizing the allowed use of the luminaire in a wet location is mandated where the luminaire will be subjected to shower spray.

E4103.3

Disconnecting Means for Pool Utilization Equipment

CHANGE TYPE. Modification

CHANGE SUMMARY. The disconnecting means for swimming pool, spa, and hot tub utilization equipment must now be located so that it is readily accessible.

2006 CODE: E4103.3 Disconnecting Means. ~~An accessible disconnecting~~ One or more means to disconnect all ungrounded conductors for all utilization equipment, other than lighting, shall be provided. Each of such means shall be readily accessible and ~~located~~ within sight from ~~all pools, spas, and hot tub equipment, and shall be located not less than 5 feet (1524 mm) from the inside walls of the pool, spa or hot tub~~ the equipment it serves.

CHANGE SIGNIFICANCE. A means for disconnecting all ungrounded conductors for pool utilization equipment is required to be located within sight of the equipment it serves. Previously, the disconnecting means had to be located only in an accessible position. Revised language now mandates that the means be readily accessible, providing for a greater degree of access. By definition, the disconnecting means must be "capable of being reached quickly, without requiring those to whom ready access is requisite to climb over or remove obstacles or to resort to portable ladders, chairs, etc."

An additional change eliminates the requirement that the disconnecting means be located at least 5 feet from the edge of the swimming pool, spa, or hot tub served by the equipment.

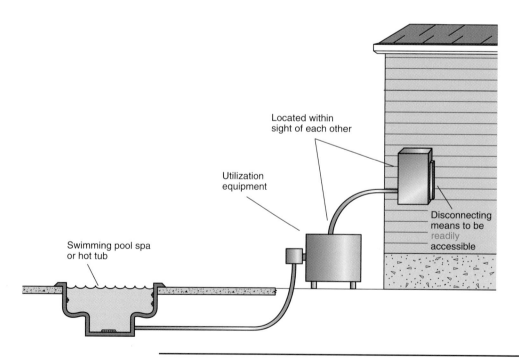

Located within sight of each other

Utilization equipment

Disconnecting means to be readily accessible

Swimming pool spa or hot tub

E4106.5.1

Servicing of Wet-Niche Luminaires

CHANGE TYPE.   Addition

CHANGE SUMMARY.   In order to provide for safe conditions during the servicing of wet-niche luminaires, it is now mandated that such luminaires be located to allow for the servicing to be done from the pool deck or another approved dry location.

2006 CODE: E4106.5.1 Servicing.   All luminaires shall be removable from the water for relamping or normal maintenance. Luminaires shall be installed in such a manner that personnel can reach the luminaire for relamping, maintenance, or inspection while on the deck or equivalently dry location.

CHANGE SIGNIFICANCE.   Wet-niche luminaires installed in a swimming pool, spa, or similar element must be capable of being removed from the water to allow for maintenance or lamp replacement. In order to eliminate the extreme shock hazard possible during service operations, it must be possible for service personnel to work from a dry location, such as the pool or spa deck.

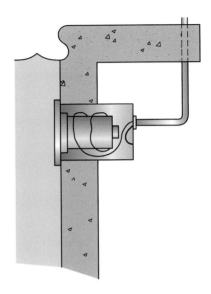

Index

A

Above-ground drainage and vent pipe, 249–251
Accessory structure, 8–9
Administration, 1–6
  schedule of permit fees, 4
  scope, 2–3
  use and occupancy, 5–6
Air conditioning equipment outlets, 281–282
Air, recirculation of, 213–214
Alarm systems, household fire, 58–59
Alarms, smoke, 58–59, 60
Alternate attachments, 110–111
Anchorage in Seismic Design Categories C, D_0, D_1,
    and D_2, 78–79
Anchorage, wall, 125–126
Appliance access
  for inspection, 204–205
  for repair, 204–205
  for replacement, 204–205
  for service, 204–205
Appliances
  in attics, 206–207
  connecting, 232
  under floor, 206–207
Approval, valve, 230
Arc-fault circuit-interruptor protection, 285–286
Area, flue, 186–187
Arms, trap, 265–266
Assemblies, attic, 171–172
Assemblies, roof, 16, 173–185
  attachment of asphalt shingles, 175–176
  hail exposure, 173–174
  hail exposure map, 173–174
  ice barriers, 177–178
  material standards, 182–183
  metal roof coverings standards, 179–181
  moderate hail exposure, 173–174
  recovering versus replacement, 184–185
  severe hail exposure, 173–174
Attachments
  alternate, 110–111
  asphalt shingles, 175–176
Attic assemblies, conditioned, 171–172
Attics, appliances in, 206–207
Auxiliary and secondary drain system,
    209–210
Auxiliary drain pan, 211–212, 224–225

B

Backers, gypsum, 143–144
Barriers
  ice, 177–178
  water-resistive, 153–154
Bathroom and shower areas, luminaires in, 292–293
Bathroom receptacles, 280
Bathtub and whirlpool bathtub valves, 239
Bearing walls, exterior, 94–96
Bedrooms, arc-fault circuit-interruptor protection
    in, 285–286
Blown or sprayed roof/ceiling insulation, building,
    193–194
Board, gypsum, 141–142
Braced wall lines and connections, framing at,
    92–93
Braced wall panels, alternate, 116–118
Bracing in Seismic Design Categories D_0, D_1, and D_2,
    122–124
Bracing wall panel adjacent to door or window
    openings, 119–121
Branch circuit and feeder requirements, 268,
    270–271
  conductor sizing of Type NM cable, 270
  panelboard circuit identification, 271
Building blown or sprayed roof/ceiling insulation,
    193–194
Building drainage and vent pipe, underground,
    249–251
Building planning, 16, 20–70
  design criteria, 20–21
  determination of design flood elevations, 69–70
  emergency escape and rescue openings, 45–46
  emergency openings under decks and porches,
    49–50
  exterior wall and opening protection, 32–34
  fire separation of two-family dwellings, 61–62
  glazing adjacent to stairways and landings, 41–42
  glazing materials permitted in hazardous
    locations, 39–40
  guards at elevated ramps, 57
  household fire alarm systems, 58–59
  irregular buildings, 27–28
  landings at exterior doors, 51–52
  landings at garage stairways, 53–54
  light activation at stairways, 35–36
  maximum slope of ramps, 55–56

Building planning, *Continued*
minimum height of sloped ceilings, 37–38
minimum uniformly distributed live loads, 29–31
operation of emergency escape and rescue openings, 47–48
protection against subterranean termites, 66–68
protection of glued-laminated members against decay, 65
protection of openings, 22–24
protection of wood members against decay, 63–64
seismic provisions, 25–26
separation of detached garage from dwelling, 43–44
smoke alarms, 58–59
smoke alarms in existing dwellings, 60
Building planning and construction, 16–187
building planning, 16
chimneys and fireplaces, 16
floors, 16
foundations, 16
roof assemblies, 16
roof-ceiling construction, 16
wall construction, 16
wall covering, 16
Building sewer pipe, 249–251
Building thermal envelope insulation, 193–194
Buildings. *See also* Dwellings
irregular, 27–28

C

Cables
conductor sizing of Type NM, 270
exposed to direct sunlight, 275
parallel to furring strips, 272–273
Categories C, D_0, D_1, and D_2, foundation anchorage in Seismic Design, 78–79
Categories D_0, D_1, and D_2, bracing in Seismic Design, 122–124
Ceiling construction. *See* Roof-ceiling construction
Ceiling insulation. *See* Roof/ceiling insulation
Ceilings
horizontal gypsum board diaphragm, 141–142
joist and rafter connections, 163–165
sloped, 37–38
Cement gypsum backers, 143–144
Chimneys and fireplaces, 16, 186–187
flue area (masonry fireplace), 186–187
Circuit, branch, 268, 270–271
panelboard circuit identification, 271
Circuit-interruptor protection
arc-fault, 285–286
ground-fault, 283–284

Circuit, panelboard, 271
Circuit receptacle outlets, small-appliance, 276–277
Compartments, shower, 236–237
Compressive or shifting soil, 73–74
Concrete and masonry foundation
dampproofing, 85–86
walls, 80–83
waterproofing, 87–88
Concrete, minimum specified compressive strength of, 75–76
Concrete slabs, under, 246–247
Conditions, seismic, 77
Conductor sizing of Type NM cable, 270
Conductors and cables exposed to direct sunlight, 275
Connecting appliances, 232
Connections
ceiling joist and rafter, 163–165
framing at braced wall lines and, 92–93
joints and, 252–253
vent, 261–262
Conservation, energy, 189–200
energy efficiency, 189–200
Construction, building planning and, 16–187
Construction, roof-ceiling, 16, 162–172
applicability limits wood truss design, 169–170
ceiling joist and rafter connections, 163–165
conditioned attic assemblies, 171–172
rafter spans for common lumber species, 166–168
structural log members, 162
Construction, wall, 16, 106–140
alternate attachments, 110–111
alternate braced wall panels, 116–118
bracing in Seismic Design Categories D_0, D_1, and D_2, 122–124
bracing wall panel adjacent to door or window openings, 119–121
corbelled masonry, 129–130
drilling and notching of top plate, 114–115
fastener schedule for structural members, 106–109
fenestration testing and labeling, 138
general window installation instructions, 133
load-bearing walls, 127–128
mullions, 139–140
requirements for ICF walls, 131–132
top plate, 112–113
wall anchorage, 125–126
wind-borne debris protection, 136–137
window sills, 134–135
Control, thermal expansion, 240–241
Control valves, shower, 238

Cooling equipment, heating and, 202, 209–212
    auxiliary and secondary drain system, 209–210
    auxiliary drain pan, 211–212
    water level monitoring devices, 209–210
Corbelled masonry, 129–130
Counters, island, 278–279
Covering, wall, 16
    cement gypsum backers, 143–144
    fiber-cement gypsum backers, 143–144
    flashing, 160–161
    general draining exterior wall assemblies,
      145–146
    glass mat gypsum backers, 143–144
    horizontal gypsum board diaphragm ceilings,
      141–142
    stone and masonry veneer, 155–159
    water-resistive barriers, 153–154
    water-resistive barrier, 147–152
    weather-resistant siding attachment, 147–152
Coverings, metal roof, 179–181
Coverings, wall, 141–161
Crawl space, openings for unvented, 89–91

D

Damage, physical, 227–228
Dampproofing, concrete and masonry foundation,
    85–86
Debris protection, wind-borne, 137–137
Decay
    protection of glued-laminated members against, 65
    protection of wood members against, 63–64
Decks and porches, emergency openings under,
    49–50
Definitions, 7–14
    accessory structure, 8–9
    approved, 10
    exterior wall, 11–12
    fire-separation distance, 13–14
Detached garage from dwelling, separation of,
    43–44
Devices and lighting fixtures, 268, 291–293
    luminaires in bathroom and shower areas,
      292–293
    receptacles in wet locations, 291
Devices, water level monitoring, 209–210
Direct sunlight, conductors and cables exposed
    to, 275
Discharge, outdoor, 213–214
Distribution, power and lighting, 268
Door or window openings, bracing wall panel
    adjacent to, 119–121
Doors, exterior, 51–52
Drain pan, auxiliary, 211–212, 224–225

Drain system, auxiliary and secondary,
    209–210
Drainage
    foundations, 71–72
    underground building, 249–251
Drainage and vent pipe, above-ground, 249–251
Drainage, sanitary, 234, 249–253
    building sewer pipe, 249–251
    joints and connections, 252–253
    pipe fittings, 249–251
    underground building drainage and vent pipe,
      249–251
Draining exterior wall assemblies, general,
    145–146
Drains, fixture, 257–258
Drywalls, repairing at panelboards, 289–290
Duct length, 215–216
Duct systems, 202, 217
    joints and seams, 217
Dwellings
    existing, 60
    separation of detached garage from, 43–44
    two-family, 61–62

E

Efficiency, energy, 189–200
    building blown or sprayed roof/ceiling insulation,
      193–194
    building thermal envelope insulation,
      193–194
    certificate, 195–196
    compliance, 191–192
    insulation and fenestration criteria,
      197–198
    mass walls, 199–200
    total UA alternative, 197–198
    U-factor alternative, 199–200
Electrical, 268–295
    branch circuit and feeder requirements, 268
    devices and lighting fixtures, 268
    power and lighting distribution, 268
    swimming pools, 268
    wiring methods, 268
Elevated ramps, guards at, 57
Elevations, flood, 69–70
Emergency escape and rescue openings, 45–46,
    47–48
Emergency openings under decks and porches,
    49–50
Enclosure support using nails and screws, 287
Enclosures in wet locations, 288
Energy conservation, 189–200
    energy efficiency, 189–200

Energy efficiency, 189–200
  building blown or sprayed roof/ceiling insulation, 193–194
  building thermal envelope insulation, 193–194
  certificate, 195–196
  compliance, 191–192
  insulation and fenestration criteria, 197–198
  mass walls, 199–200
  total UA alternative, 197–198
  U-factor alternative, 199–200
Equipment
  heating and cooling, 202, 209–212
  pool utilization, 294
Equipment outlets
  air conditioning, 281–282
  heating, 281–282
  refrigeration, 281–282
Escape and rescue openings, emergency, 45–46, 47–48
Exhaust systems, 202, 213–216
  duct length, 215–216
  outdoor discharge, 213–214
  recirculation of air, 213–214
Existing dwellings, smoke alarms in, 60
Exposure
  hail, 173–174
  moderate hail, 173–174
  severe hail, 173–174
Exposure map, hail, 173–174
Extensions
  roof, 256
Exterior bearing walls, girder spans and header spans for, 94–96
Exterior doors, landings at, 51–52
Exterior wall, 11–12
  assemblies, 145–146
  and opening protection, 32–34

F

Fastener schedule for structural members, 106–109
Feeder requirements, branch circuit and, 268
Fees, permit, 4
Fenestration criteria, insulation and, 197–198
Fenestration testing and labeling, 138
Fiber-cement gypsum backers, 143–144
Fire alarm systems, household, 58–59
Fireblocking required, 97
Fire-separation distance, 13–14
Fire separation of two-family dwellings, 61–62
Fireplaces, chimneys and, 16, 186–187
  flue area (masonry fireplace), 186–187
Fireplaces, masonry, 186–187
Fittings, pipe, 249–251

Fixture drains, 257–258
Fixtures
  lighting, 268, 291–293
Fixtures per trap, number of, 263–264
Fixtures, plumbing, 234, 236–239, 265–266
  bathtub and whirlpool bathtub valves, 239
  shower compartment general and access, 236–237
  shower control valves, 238
Flashing, 160–161
Flood elevations, determination of design, 69–70
Floor, appliances under, 206–207
Floor framing, limits for steel, 101–102
Floor sheathing. *See* Subfloor sheathing
Floor underlayment. *See* Subfloor underlayment
Floor ventilation. *See* Underfloor ventilation
Floors, 16, 92–105
  allowable joist spans, 103–104
  allowable spans and loads for wood structural panels, 98–100
  combination subfloor underlayment, 98–100
  fireblocking required, 97
  framing at braced wall lines and connections, 92–93
  girder spans and header spans for exterior bearing walls, 94–96
  limits for steel floor framing, 101–102
  reinforcement support, 105
  roof and subfloor sheathing, 98–100
  solid, 229
  trusses, 103–104
Flue area (masonry fireplace), 186–187
Foundation walls, concrete and masonry, 80–83
Foundations, 16, 71–91
  anchorage in seismic design categories C, D_0, D_1, and D_2, 78–79
  compressive or shifting soil, 73–74
  concrete and masonry, 85–86, 87–88
  concrete and masonry foundation dampproofing, 85–86
  concrete and masonry foundation walls, 80–83
  concrete and masonry foundation waterproofing, 87–88
  drainage, 71–72
  minimum specified compressive strength of concrete, 75–76
  openings for under-floor ventilation, 89–91
  openings for unvented crawl space, 89–91
  retaining walls, 84
  seismic conditions, 77
  and supports, 208
Framing at braced wall lines and connections, 92–93
Framing, steel floor, 101–102

Fuel gas, 221–232
  auxiliary drain pan, 224–225
  breather and relief, 231
  connecting appliances, 232
  definitions, 222
  listed and labeled, 223
  manual gas valve standards, 230
  piping in solid floors, 229
  point of delivery, 222
  prohibited locations for gas piping, 226
  protection against physical damage, 227–228
  valve approval, 230
  vent piping, 231
  venting of regulators, 231
Furring strips, protection of cables parallel to, 272–273

G

Garage, separation of detached, 43–44
Garage stairways, landings at, 53–54
Gas, fuel, 221–232
  auxiliary drain pan, 224–225
  breather and relief, 231
  connecting appliances, 232
  definitions, 222
  listed and labeled, 223
  manual gas valve standards, 230
  piping in solid floors, 229
  point of delivery, 222
  prohibited locations for gas piping, 226
  protection against physical damage, 227–228
  valve approval, 230
  vent piping, 231
  venting of regulators, 231
Gas piping, prohibited locations for, 226
Gas valve standards, manual, 230
Girder spans for exterior bearing walls, 94–96
Glass mat gypsum backers, 143–144
Glazing
  adjacent to stairways and landings, 41–42
  materials permitted in hazardous locations, 39–40
Glued-laminated members against decay, protection of, 65
Ground-fault circuit-interruptor protection, 283–284
Guards at elevated ramps, 57
Gypsum backers
  cement, 143–144
  fiber-cement, 143–144
  glass mat, 143–144
Gypsum board diaphragm ceilings, horizontal, 141–142

H

Hail exposure, 173–174
  map, 173–174
  moderate, 173–174
  severe, 173–174
Hanger spacing intervals, 218–220
Hazardous locations, glazing materials permitted in, 39–40
Header spans for exterior bearing walls, 94–96
Heating and cooling equipment, 202, 209–212
  auxiliary and secondary drain system, 209–210
  auxiliary drain pan, 211–212
  water level monitoring devices, 209–210
Heating equipment outlets, 281–282
Height of sloped ceilings, minimum, 37–38
Horizontal gypsum board diaphragm ceilings, 141–142
Household fire alarm systems, 58–59
Hydronic piping, 202, 218–220
  hanger spacing intervals, 218–220
  materials, 218–220
  piping joints, 218–220
  piping joints for low-temperature piping, 218–220
  piping materials, 218–220

I

I-Codes, 7
IBC. *See International Building Code*
IBHS. *See* Institute for Business and Home Safety
ICC. *See* International Code Council
Ice barriers, 177–178
ICF walls, requirements for, 131–132
IEBC. *See* International Existing Building Code
IECC. *See* International Energy Conservation Code
IFGC. *See* International Fuel Gas Code
Installation
  polyethylene plastic piping, 242–243
  window, 133
Institute for Business and Home Safety (IBHS), 173
Insulation
  building blown or sprayed roof/ceiling, 193–194
  building thermal envelope, 193–194
  and fenestration criteria, 197–198
International Building Code (IBC), 1, 3
International Code Council (ICC), 7
International Energy Conservation Code (IECC), 189
International Existing Building Code (IEBC), 2, 3
International Fuel Gas Code (IFGC), 221
International Residential Code (IRC), 1, 2
IRC. *See International Residential Code*

Irregular buildings, 27–28
Island counter spaces, receptacle outlets at wall and, 278–279

J

Joints
    and connections, 252–253
    piping, 218–220
    and seams, 217
Joist spans, allowable, 103–104
Joists, ceiling, 163–165

L

Labeling, fenestration testing and, 138
Laminated. *See* Glued-laminated
Landings
    at exterior doors, 51–52
    at garage stairways, 53–54
    glazing adjacent to stairways and, 41–42
Laundry and utility sinks, 283–284
Light activation at stairways, 35–36
Lighting distribution, power and, 268, 276–290
    air conditioning equipment outlets, 281–282
    arc-fault circuit-interruptor protection in
        bedrooms, 285–286
    bathroom receptacles, 280
    enclosure support using nails and screws, 287
    enclosures in wet locations, 288
    ground-fault circuit-interruptor protection,
        283–284
    heating equipment outlets, 281–282
    laundry and utility sinks, 283–284
    receptacle outlets at wall and island counter
        spaces, 278–279
    refrigeration equipment outlets, 281–282
    repairing drywall at panelboards, 289–290
    small-appliance circuit receptacle outlets,
        276–277
Lighting fixtures, devices and, 268, 291–293
    luminaires in bathroom and shower areas,
        292–293
    receptacles in wet locations, 291
Lines and connections, framing at braced wall, 92–93
Live loads, minimum uniformly distributed, 29–31
Load-bearing walls, 127–128
Loads for wood structural panels, allowable spans
    and, 98–100
Loads, live, 29–31
Locations
    hazardous, 39–40
    wet, 288, 291

Locations for gas piping, prohibited, 226
Log members, structural, 162
Low-temperature piping, piping joints for,
    218–220
Lumber species, rafter spans for common,
    166–168
Luminaires
    in bathroom and shower areas, 292–293
    servicing of wet-niche, 295

M

Manual gas valve standards, 230
Maps, hail exposure, 173–174
Masonry
    corbelled, 129–130
    fireplace, 186–187
Masonry foundation dampproofing, concrete and,
    85–86
Masonry foundation walls, concrete and, 80–83
Masonry foundation waterproofing, concrete and,
    87–88
Masonry veneer, stone and, 155–159
Mass walls, 199–200
Material standards (roof assemblies), 182–183
Materials
    glazing, 39–40
    piping, 218–220
Mechanical, 202–220
    duct systems, 202
    exhaust systems, 202
    general mechanical system requirements, 202
    heating and cooling equipment, 202
    hydronic piping, 202
Mechanical system requirements, general, 202,
    204–208
    appliance access for inspection, repair, and
        replacement, 204–205
    appliances in attics, 206–207
    appliances under floor, 206–207
    foundations and supports, 208
Members, structural, 106–109
Metal roof coverings standards, 179–181
Methods, wiring, 268, 272–275
    conductors and cables exposed to direct sunlight,
        275
    protecting Type NM cable from physical damage,
        274
    protection of cables parallel to furring strips,
        272–273
Moderate hail exposure, 173–174
Monitoring devices, water level, 209–210
Mullions, 139–140

N

Nails and screws, enclosure support using, 287
National Electrical Code (NEC), 268
National Fire Protection Association (NFPA), 268
NEC. *See* National Electrical Code
NFPA. *See* National Fire Protection Association
NM cable, conductor sizing of Type, 270
NM cable from physical damage, protecting Type, 274

O

Occupancy, use and, 5–6
Opening protection, exterior wall and, 32–34
Openings
    emergency escape and rescue, 45–46
    operation of emergency escape and rescue, 47–48
    protection of, 22–24
    for under-floor ventilation, 89–91
    for unvented crawl space, 89–91
Openings, door or window, 119–121
Openings under decks and porches, emergency, 49–50
Outdoor discharge, 213–214
Outlets
    air conditioning equipment, 281–282
    heating equipment, 281–282
    refrigeration equipment, 281–282
    small-appliance circuit receptacle, 276–277

P

Panelboards
    and circuit identification, 271
    repairing drywalls at, 289–290
Panels
    allowable spans and loads for wood structural, 98–100
    wall, 116–118, 119–121
Pans
    auxiliary drain, 224–225
    drain, 211–212
Permit fees, schedule of, 4
Physical damage
    protecting Type NM cable from, 274
    protection against, 227–228
Pipe fittings, 249–251
Pipes
    building sewer, 249–251
    vent, 249–251
    water service, 244–245

Piping
    gas, 226
    joints for low-temperature, 218–220
    polyethylene plastic, 242–243
    in solid floors, 229
    vent, 231
Piping, hydronic, 202, 218–220
    hanger spacing intervals, 218–220
    materials, 218–220
    piping joints, 218–220
    piping joints for low-temperature piping, 218–220
    piping materials, 218–220
Piping joints, 218–220
    for low-temperature piping, 218–220
Piping materials, 218–220
Planning, building, 16, 16–187, 20–70
    design criteria, 20–21
    determination of design flood elevations, 69–70
    emergency escape and rescue openings, 45–46
    emergency openings under decks and porches, 49–50
    exterior wall and opening protection, 32–34
    fire separation of two-family dwellings, 61–62
    glazing adjacent to stairways and landings, 41–42
    glazing materials permitted in hazardous locations, 39–40
    guards at elevated ramps, 57
    household fire alarm systems, 58–59
    irregular buildings, 27–28
    landings at exterior doors, 51–52
    landings at garage stairways, 53–54
    light activation at stairways, 35–36
    maximum slope of ramps, 55–56
    minimum height of sloped ceilings, 37–38
    minimum uniformly distributed live loads, 29–31
    operation of emergency escape and rescue openings, 47–48
    protection against subterranean termites, 66–68
    protection of glued-laminated members against decay, 65
    protection of openings, 22–24
    protection of wood members against decay, 63–64
    seismic provisions, 25–26
    separation of detached garage from dwelling, 43–44
    smoke alarms, 58–59
    smoke alarms in existing dwellings, 60
Plastic piping installation, polyethylene, 242–243
Plastic, polyethylene, 248
Plate, top, 112–113, 114–115
Plumbing, 234–266
    fixtures, 234
    sanitary drainage, 234

Plumbing *Continued*
  traps, 234
  vents, 234
  water supply and distribution, 234
Plumbing fixtures, 236–239, 245–255
  bathtub and whirlpool bathtub valves, 239
  shower compartment general and access, 236–237
  shower control valves, 238
Polyethylene plastic, 248
  piping installation, 242–243
Pool utilization equipment, disconnecting means for, 294
Pools, swimming, 268, 294–295
  disconnecting means for pool utilization equipment, 294
  servicing of wet-niche luminaires, 295
Porches, emergency openings under decks and, 49–50
Power and lighting distribution, 268, 276–290
  air conditioning equipment outlets, 281–282
  arc-fault circuit-interruptor protection in bedrooms, 285–286
  bathroom receptacles, 280
  enclosure support using nails and screws, 287
  enclosures in wet locations, 288
  ground-fault circuit-interruptor protection, 283–284
  heating equipment outlets, 281–282
  laundry and utility sinks, 283–284
  receptacle outlets at wall and island counter spaces, 278–279
  refrigeration equipment outlets, 281–282
  repairing drywall at panelboards, 289–290
  small-appliance circuit receptacle outlets, 276–277
Protection
  of cables parallel to furring strips, 272–273
  exterior wall and opening, 32–34
  of glued-laminated members against decay, 65
  ground-fault circuit-interruptor, 283–284
  of openings, 22–24
  against physical damage, 227–228
  against subterranean termites, 66–68
  wind-borne debris, 136–137
  of wood members against decay, 63–64
Protection in bedrooms, arc-fault circuit-interruptor, 285–286
Provisions, seismic, 25–26

R

Rafter connections, ceiling joist and, 163–165
Rafter spans for common lumber species, 166–168

Ramps
  elevated, 57
  maximum slope of, 55–56
Receptacle outlets
  small-appliance circuit, 276–277
  at wall and island counter spaces, 278–279
Receptacles
  bathroom, 280
  in wet locations, 291
Recirculation of air, 213–214
Recovering versus replacement (roof assemblies), 184–185
Refrigeration equipment outlets, 281–282
Regulators, venting of, 231
Reinforcement support, 105
Replacement versus recovering (roof assemblies), 184–185
Rescue openings
  emergency escape and, 45–46
  operation of emergency escape and, 47–48
Retaining walls, 84
Roof and subfloor sheathing, 98–100
Roof assemblies, 16, 173–185
  attachment of asphalt shingles, 175–176
  hail exposure, 173–174
  hail exposure map, 173–174
  ice barriers, 177–178
  material standards, 182–183
  metal roof coverings standards, 179–181
  moderate hail exposure, 173–174
  recovering versus replacement, 184–185
  severe hail exposure, 173–174
Roof-ceiling construction, 16, 162–172
  applicability limits wood truss design, 169–170
  ceiling joist and rafter connections, 163–165
  conditioned attic assemblies, 171–172
  rafter spans for common lumber species, 166–168
  structural log members, 162
Roof/ceiling insulation, building blown or sprayed, 193–194
Roof coverings standards, metal, 179–181
Roof extension, 256

S

Sanitary drainage, 234, 249–253
  above-ground drainage and vent pipe, 249–251
  building sewer pipe, 249–251
  joints and connections, 252–253
  pipe fittings, 249–251
  underground building drainage and vent pipe, 249–251
Schedule of permit fees, 4
Screws, enclosure support using nails and, 287

Seams, joints and, 217
Secondary drain system, auxiliary and, 209–210
Seismic
    conditions, 77
    provisions, 25–26
Seismic Design Categories
    C, D_0, D_1, and D_2, 78–79
    D_0, D_1, and D_2, 122–124
Severe hail exposure, 173–174
Sewer pipe, building, 249–251
Sheathing, roof and subfloor, 98–100
Shifting soil, compressive or, 73–74
Shower
    compartment general and access, 236–237
    control valves, 238
Shower areas, luminaires in bathroom and, 292–293
Siding attachment and minimum thickness,
    147–152
Sills, window, 134–135
Sinks, laundry and utility, 283–284
Size of traps and trap arms, 265–266
Slabs, under concrete, 246–247
Slope of ramps, maximum, 55–56
Sloped ceilings, minimum height of, 37–38
Small-appliance circuit receptacle outlets,
    276–277
Smoke alarms, 58–59
    in existing dwellings, 60
Soil, compressive or shifting, 73–74
Solid floors, piping in, 229
Space, unvented crawl, 89–91
Spans
    allowable joist, 103–104
    girder, 94–96
    header, 94–96
    and loads for wood structural panels, 98–100
    rafter, 166–168
Species, lumber, 166–168
Sprayed roof/ceiling insulation, building blown or,
    193–194
Stairways
    garage, 53–54
    and landings, 41–42
    light activation at, 35–36
Standards, manual gas valve, 230
Steel floor framing, limits for, 101–102
Stone and masonry veneer, 155–159
Strips, furring, 272–273
Structural log members, 162
Structural members, fastener schedule for,
    106–109
Structure, accessory, 8–9
Subfloor sheathing, roof and, 98–100
Subfloor underlayment, combination, 98–100

Subterranean termites, protection against, 66–68
Sunlight, direct, 275
Supply, water, 234
Supports
    enclosure, 287
    foundations and, 208
    reinforcement, 105
Swimming pools, 268, 294–295
    disconnecting means for pool utilization
        equipment, 294
    servicing of wet-niche luminaires, 295
Systems
    auxiliary and secondary drain, 209–210
    household fire alarm, 58–59
    mechanical, 202, 204–208
Systems, duct, 202, 217
    joints and seams, 217
Systems, exhaust, 202, 213–216
    duct length, 215–216
    outdoor discharge, 213–214
    recirculation of air, 213–214

T

Termites, subterranean, 66–68
Testing and labeling, fenestration, 138
Thermal envelope insulation, building, 193–194
Thermal expansion control, 240–241
Top plate, 112–113
    drilling and notching of, 114–115
Trap arms, size of traps and, 265–266
Traps, 234, 263–266
    number of fixtures per, 263–264
    size of traps and trap arms for plumbing fixtures,
        265–266
Traps and trap arms, size of, 265–266
Truss design, wood, 169–170
Trusses, floor, 103–104
Two-family dwellings, fire separation of, 61–62
Type NM cable
    conductor sizing of, 270
    protecting from physical damage, 274

U

U-factor alternative, 199–200
UA alternative, total, 197–198
Under concrete slabs, 246–247
Under-floor ventilation, openings for, 89–91
Underground building drainage and vent pipe,
    249–251
Unvented crawl space, openings for, 89–91
Use and occupancy, 5–6
Utility sinks, laundry and, 283–284

V

Valves
  approval of, 230
  bathtub and whirlpool bathtub, 239
  gas, 230
  shower control, 238
Veneer, stone and masonry, 155–159
Vent connections and size, 261–262
Vent pipes
  above-ground drainage and, 249–251
  underground building drainage and, 249–251
Vent piping, 231
Vent stacks and stack vents, 254–255
Vents, 234, 254–262
  fixture drains, 257–258
  roof extension, 256
  vent connections and size, 261–262
  vent stacks and stack vents, 254–255
  wet vent permitted, 259–260
Ventilation, openings for under-floor, 89–91

W

Wall anchorage, 125–126
Wall and island counter spaces, 278–279
Wall construction, 16, 106–140
  alternate attachments, 110–111
  alternate braced wall panels, 116–118
  bracing in Seismic Design Categories D_0, D_1, and D_2, 122–124
  bracing wall panel adjacent to door or window openings, 119–121
  corbelled masonry, 129–130
  drilling and notching of top plate, 114–115
  fastener schedule for structural members, 106–109
  fenestration testing and labeling, 138
  general window installation instructions, 133
  load-bearing walls, 127–128
  mullions, 139–140
  requirements for ICF walls, 131–132
  top plate, 112–113
  wall anchorage, 125–126
  wind-borne debris protection, 136–137
  window sills, 134–135
Wall coverings, 16, 141–161
  cement gypsum backers, 143–144
  fiber-cement gypsum backers, 143–144
  flashing, 160–161
  general draining exterior wall assemblies, 145–146
  glass mat gypsum backers, 143–144
  horizontal gypsum board diaphragm ceilings, 141–142

  stone and masonry veneer, 155–159
  water-resistive barrier, 147–152, 153–154
  weather-resistant siding attachment, 147–152
Wall lines and connections, framing at braced, 92–93
Wall panel adjacent to door or window openings, 119–121
Wall panels, alternate braced, 116–118
Walls. *See also* Drywalls
  concrete and masonry foundation, 80–83
  exterior, 11–12, 32–34, 145–146
  girder spans for exterior bearing, 94–96
  header spans for exterior bearing, 94–96
  ICF, 131–132
  load-bearing, 127–128
  mass, 199–200
  retaining, 84
Water level monitoring devices, 209–210
Water service pipe, 244–245
Water supply and distribution, 234, 240–248
  under concrete slabs, 246–247
  polyethylene plastic, 248
  polyethylene plastic piping installation, 242–243
  thermal expansion control, 240–241
  water service pipe, 244–245
Waterproofing, concrete and masonry foundation, 87–88
Water-resistive barriers, 153–154
Weather-resistant
  siding attachment, 147–152
Wet locations
  enclosures in, 288
  receptacles in, 291
Wet-niche luminaires, servicing of, 295
Wet vent permitted, 259–260
Whirlpool bathtub valves, bathtub and, 239
Wind-borne debris protection, 136–137
Window installation instructions, general, 133
Window openings, bracing wall panel adjacent to door or, 119–121
Window sills, 134–135
Wiring methods, 268, 272–275
  conductors and cables exposed to direct sunlight, 275
  protecting Type NM cable from physical damage, 274
  protection of cables parallel to furring strips, 272–273
Wood members against decay, protection of, 63–64
Wood structural panels, allowable spans and loads for, 98–100
Wood truss design, applicability limits, 169–170